Revolutionary Wisdom

Organic Psychology in Action

How to invoke your fifty-four sense natural intelligence and improve any relationship.

Michael J. Cohen

Stacey S. Mallory

Heart-Centered Applied Ecopsychology-Ecotherapy-Whole Life Thinking

Learn how to sense and feel like the joy of a natural area works.

Project NatureConnect
Preliminary Edition

CreateSpace.com for Project NatureConnect
Warrantied GreenWave-54 whole life science and relationships.
P. O. Box 1605 Friday Harbor WA 98250 (360) 378-6313
www.ecopsych.com | nature@interisland.net

Happily, the art and science of Organic Psychology, in this textbook discovers the source of, and remedy for, our runaway personal, social and environmental deterioration. The book itself is the key unifying tool needed to achieve this goal. Many dysfunctions and disorders that accompany our great technological achievements indicate that, unlike the life of nature, we reward ourselves for producing and suffering a monumental short circuit in the way Industrial Society works. As we addict to our increasingly advanced artificial lifestyles, we increase our loss of Nature's self-correcting ways. Our artifacts can't replace the latter's wisdom and healthy satisfactions, so we excessively want and artificially satisfy ourselves further reducing our wellness. With respect to the life of Nature and its optimums of harmonic balance, purity, and beauty, **there is no substitute for the real thing.** Learning how to **tap into and enjoy the 54-senses consciously, the organic intelligence of our planet's life in natural areas improves our relationships.** *Revolutionary Wisdom* enables each of us, at any moment, to stop our madness by consciously blending our lives with the life of Nature/Earth and each other as fiduciaries and transforms those who withhold this remedy from us. This book welcomes and empowers people to integrate its methods and materials.

AUTHORS

Michael J. Cohen, Ed.D., Ph.D.

While developing the GreenWave-54 process since 1953, Mike has achieved several Master's and Doctoral degrees, written 11 books, directed sensory organic university environmental education and outdoor education courses and degree programs for over 50 years as well as, for decades, developed and lived outdoors on year-long, "utopian community," environmental education expeditions. He is recognized as a Maverick Genius and has received the Distinguished World Citizen award for his contributions to nature-connected counseling and healing. He conceived the 1985 symposium 'Is the Earth a Living Organism,' and his work consists of full, self-evident facts of life that include our 54 natural senses. Although scrutinized since 1982, these 54 facts of life have yet to be denied.

Stacey S. Mallory, Ph.D.

As the Director of Project NatureConnect along with her pioneering work in Eco-Art Therapy her background demonstrates, Dr. Mallory thinks "outside" the box. She is certified in Eco-Art Therapy. She has various Diplomas, accredited by CPD, in the Art Therapy. As well as one in Reiki and Color Therapy and Rational Emotive Behavior Therapy (REBT). She is faculty at Akamai University and Portland State University where she teaches Applied Ecopsychology.

Contents

Dedication

In loving memory of Joey "Faulkner-Mallory" who was cruelly shot and killed for the fun or sport of it on September 13, 2017. Joey was our loyal therapy dog companion, a friend who helped many people heal through Organic Psychology and Eco-Art Therapy.

We dedicate this book to all plants and animals, including people, that suffer from our destructive disconnection from Nature to those who want to replace its abusive effects by reconnecting with the life of their extraordinary, self-correcting roots in natural areas.

Joey did not die in vain.

On September 15, 2017, the Faulkner-Mallory family received this message from Joey by email. It brought them to tears because it conveys the core of Organic Psychology. With death, as with life, the family continues to teach its truth.

> Do not stand at my grave and weep
> I am not there. I do not sleep.
> I am a thousand winds that blow.
> I am the diamond glints on snow.
> I am the sunlight on ripened grain.
> I am the gentle autumn rain.
> When you awaken in the morning's hush
> I am the swift uplifting rush
> Of quiet birds in circled flight.
> I am the soft stars that shine at night.
> Do not stand at my grave and cry;
> I am not there. I did not die.

Joey was shot because with respect, actually disrespect, to living as part of the whole of life, many of our contemporary ways do not make sense. This heartbreak occurs because we learn to think and relate using only eight of our fifty-four natural senses that, in and around us, blend to create Nature's balance and peace without producing any garbage. It is destructive we learn to spend 99 percent of our time out-of-tune with Nature.

The omission from our lives of forty-six of our inherent senses leaves most of us with a whole-life sensibility IQ of 15, that of an idiot who is prejudiced against nature. Along with, being out of tune with Nature 99 percent of the time helps explain our excessive personal, social and environmental disorders, including Joey's death. Like the greater-than-human life of Earth itself, Joey's life mattered because he was and, transformed, remains it.

With reverence for Joey and to remedy our anti-nature non-sense, we urge you to master and teach the fifty-four-sense revolutionary wisdom process that this book offers. It unifies and converts our prejudice against nature into love. Individuals and communities that practice it reduces violence and derangements while helping others do the same.

KNWA Northwest Arkansas News:

"He was bleeding from his neck and gasping for air. There was nothing I could do," said Crystal Faulkner whose dog was shot. In an instant, Faulkner's best friend was taken away. Faulkner said suddenly she heard a loud boom. "I glanced down at my dog, and I saw him stagger," she said. "I watched him take his last breath." The killer now on $250,000 bail faces 6 years in prison or a $10,000 fine. "People need to get the message that you can't just take your gun and shoot a dog, or a pet," said Faulkner." This is the **first** prosecution for this type of crime in nine years.

Dear Editor, Administrator, and Other Interested Parties,

Although you may find it repugnant, it is urgent, as well as your responsibility, to present unwelcome facts that give voice to the voiceless. Your profession demands that you provide complete and accurate information about the source and remedy of any issue. This includes the dilemma of how the way we learn to relate increases climate change, species extinction, mental illness, resource depletion, obesity, violence and rampant unfairness. The facts we need to remedy this catastrophe are known, and you are accountable to present them as evidence to consider.

It is undeniable that the means are available to help us reverse the devastation we continue to create, including the murder, described below. We scientifically know how to produce the misery we cause, and we scientifically know how to reverse it. Industrial Society is paying those in power, including yourself, to omit the latter while we progressively suffer. Sadly, you no doubt suffer too, but, no matter how inconvenient, your oath as a journalist is to courageously bring to light vital information and be part of the solution.

Please make the contribution only you can make. Boldly disclose the organic remedy to the Earth misery we cause. It is a core, unifying truth and process that we are indoctrinated to overlook. For example, the indisputable fact is that although you and I are strangers and many miles apart from each other, we both know and agree that you are reading these words right now and breathing, too. It is self-evident, and self-evidence is the most accurate form of organic scientific fact, it registers directly on our nervous system biology and unifies.

It is equally evident that you and I have a sense of thirst yet it is not one of the five senses we learn to value. The truth is that we have 54 natural senses that connect us to the life of Earth and its self-correcting purity. We are paid to exploit rather than heed their wisdom.

Please consider and investigate the following. Why wouldn't you want your readers to know that this 54-sense art and science is available?

Crystal Faulkner and Stacey Mallory, are both Ph.D. faculty members of Project NatureConnect's 54-sense art and science of Organic Psychology. They say its process gave them the means and fortitude to report the malicious death of Joey and obtain justice for it and all of life. Their dedication successfully produced in Arkansas the first prosecution of this type of crime in nine years. Most of the 2.988 million people of Arkansas need to learn Organic Psychology, as do we all.

We must apply Organic Psychology to reverse the catastrophe of our ever-increasing erosion of morality and human decency as we continue our torture, child abuse, racism, emotional disorders, sexism, homophobia, violence and antisemitism ad nauseam. To stop this evil, this text for nature-connecting, organics welcomes and enables individuals or organizations to benefit from its application. Presently, our experts identify the problems that injure us, but they refuse to give us the available Revolutionary Wisdom tools we need to correct them.

Information and Videos are available:

Natasha Redina[1] | Compassion Charter[2] |A reviewed Journal Article[3]

1 https://vimeo.com/211249559

2 https://www.youtube.com/watch?v=zWCs52q3H-8

3 http://www.ecopsych.com/GREENWAVEBETA.docx

Introduction

Concerning the life of Nature/Earth that is also your life, when that relationship is abused or sick what disease makes you think that your life is well? That disease denies its existence. It also contradicts its organic remedy, the art, and science of revolutionary wisdom.

Industrial Society[4] conditions us to suffer from 'Earth Misery'[5] a runaway, natural resource, species extinction and mental illness catastrophe. It deteriorates personal, social and environmental well-being by ignoring[6] its antidote. The antidote enables the life of Earth[7] to activate our 54 natural senses[8], and remedy their hurtful abuse, in organic ways reverse the nature-disconnected relationships that produce our derangements and crippling budgets.

The source of Earth Misery and most other problems are self-evident. Without producing our garbage abusiveness and disorders, the self-correcting ways of unadulterated Nature organically create optimums of life, diversity, cooperation, balance, and well-being. We suffer because our society socializes us with **stories that detach** over 98 percent of our life from the sane and healing wisdom of the life of Nature/Earth.

Five decades of Organic Psychology research in natural areas exploring "attraction being conscious of what it is attracted to" warranties that this book's, nature-connected expedition learning process validates our sensations and feelings to be clear. These facts empower the art of the scientific method **to reconnect and happily restore** our missing whole-life integrity[6]. Funded and internet available, backyard or back-country this tool helps people create extraordinary moments that let Earth teach and heal.

Things go better with Nature

Reading Revolutionary Wisdom while 54-sense, expedition, connected with authentic natural areas, backyard or backcountry, empowers us to learn, and to teach others how to build evidence-based, heart-centered relationships that include the restorative, self-correcting ways of the life of Earth and its inherent well-being.

---

[4] http://www.ecopspych.com/journalinstitution.html

[5] http://www.ecopsych.com/zombie2.html

[6] http://www.ecopsych.com/grandjury.html

[7] http://www.ecopsych.com/livingplanetearthkey.html

[8] http://www.ecopsych.com/prejudicebigotry.html

Over the eons, Nature/Earth's balance of its moment-by-moment optimums of life, diversity, beauty, purity, peace, wellness, community and cooperation has shown that it does not produce contemporary humanity's runaway predicaments. **It makes no sense for us not to correct our flawed processes that cause our troubles.** We must establish adequate time and space for our fifty-four natural senses to connect with Nature so we may beneficially think, relate and blend with its wisdom. This includes transforming some or our established procedures into **more appropriately rewarding relationships**.

This book works because its Organic Psychology scientifically helps us resolve problems and create well-being in our technology-attached lives. It achieves this by holistically applying our civilization's core truth, evidenced-based stories, which were discovered by:

> **Thales of Miletus** who successfully ***omitted the mystical and supernatural from accounts of nature.***
>
> **Pythagoras** who determined correctly that ***the Universe contained a logical, inherent mathematical order***, as demonstrated by the sequence of a seed growing into a tree.
>
> **Copernicus** who ***combined math and science to validate*** the Sun, not the Earth, was the center of the solar system.
>
> **Albert Einstein** who determined that ***the Universe consisted of a Unified Field that made its own space and time moment-by-moment,*** including this moment. (You are experiencing these words in "real universe time.")

The life of Planet Earth has always produced a limited surplus of resources that help new life relationships grow in organic balance with the planet and each other. Knowingly, since 1949, our methods and materials of math and science have used Earth's surplus in non-organic ways to expand our artificial world. As a result, in 1974 Earth ran short of the life-sustaining resources it needed by the end of that year. Since then, Earth has increasingly become more resource deficient because science and technology have **continued not to operate in organic ways, balanced ways. Neither have they** identified or corrected the source of our ever-increasing disconnection from Nature's rational balance and beauty, around and in us. Revolutionary Wisdom is an expedition that gives us the missing organic tools we need to identify the source and remedy this disorder. It helps us think and relate to the whole-life integrity of how the life of Nature and Earth work.

The art and science in this expeditionary book deserve your trust because it results from decades of first-hand, 54-sense, natural area experiences by others. It creates the opportunity for you to personally register these experiences in authentic local natural areas as you engage in this book's organic application of Ecopsychology.

Your Time Line: How to Make Nature Your Second Language

You can **read this book's information** in 4-5 hours to acquaint yourself with its benefits. You can also accomplish this in 30 minutes, an hour or less, by reading the article *An Objective Search for Mother Earth.* Either way, the information provided may influence how you think and feel. However, that seldom changes our behavior or relationships, and that is what we need to increase personal, social, and environmental well-being; locally and globally.

If you utilize **this expedition book as a course or remedy** by following its instructions, you will, in time, gain an organic relationship-building habit that you can teach and that increases wellness to the benefit of all. **This unique training may take seven months or longer,** depending on how much time and dedication you put into it. For this reason, the book occasionally repeats its main points so they won't be forgotten later on. Completed thoroughly, with a mentor and/or partner(s), this expedition book can be your key to quickly completing an Applied Ecopsychology degree, certification and livelihood[29]. They become like reading a driver's training book after you have already mastered your ability to drive a car.

The Core Metaphor for Contemporary Life and its Discontents

Industrial Society emotionally attaches us to drive an advanced technology automobile. It is not entirely organic, so it makes us produce and suffer our personal, local and global misery[5]. This is because to satisfy our practical needs and emotional frustrations we enjoy rewarding satisfactions from speeding our car down the highway.

Suddenly we see that we are going to crash into a group of families having a picnic in a beautiful natural area.

Because we have not yet learned how to fully activate the car's advanced braking and steering system, in anguish, we hope and pray that it will change direction as we fearfully scream "Oh my god" "stop" or "whoa," as if the vehicle was a runaway spirit or horse.

These reactions are unscientific and outdated. They do not stop our advanced technology automobile. For this reason, we wreak havoc on innocent people, places and things including ourselves as passengers.

> What Industrial Society's science knows but seldom teaches us is that, as in real life, every atom, energy, and relationship in this scenario, including us, was once part of the life of a natural area. There, in congress with other atoms and energies, and without using stories,

through the eons, these components were attracted to organize themselves and relate in the life of Nature/Earth's balanced, self-correcting diversity, purity, and growth. This is significant; **because of these same organically, healthy relationships are right now operating in that natural area while our runaway car is tearing it and them asunder.**

FACT: The means to connect with the life of Nature/Earth and stop the car are readily available while our socialization still teaches us to conquer and destroy it.

On many levels, we sense and are wounded by our uncontrolled car injuring us because **the life of Nature/Earth in, around and as us is identical.** Our story that we are superior and immune to or protected from the car's destructiveness is a self-inflicted delusion. It is a misguided fable that the natural world cannot directly correct because, being speechless, the life of Nature/Earth neither tells, relates with, nor understands stories. It is "dumb," and we treat it accordingly.

The Organic Challenge for Art and Science

It is common knowledge that the art of our science based contemporary lives, correctly teach us to know we are born of and share our experiences with the life of Nature and Earth.

However, scientifically, and to our loss, our stories have excessively closeted our indoor lives from the self-correcting, pure and balanced ways of Nature/Earth.

Our closeting has been done in ways that conquer and exploit the life of Nature, in and around us, in ways that unduly hurt and stress us.

We have known since 1950 that our indoor-bonded life and the natural world are deteriorating due to our estrangement from the natural world. This is because this separation harms and limits the whole of life and its self-correcting powers.

Studies determined in 1974 that the life of Earth, that year and every year since increasingly ran out of the physical and energy resources it needed to sustain the growth of our planet in balance with humanity.

Today, 2018, the life of Nature/Earth/Us suffers a 47 percent, and counting, increase in natural resource depletion, species extinction, climate change, mental illness, obesity and most other disorders. **We need the life of another Planet Earth, half Earth's size, to attach to Earth. Nobody knows where this other planet is or how to connect with it.**

No contemporary art or science has appeared that identifies no less stop and reverses the "point" catastrophic source of our personal, social and environmental undeclared war against nature that makes contemporary society normally include us suffering from the post-traumatic stress of our excessive childhood disconnectedness. This results because our traditional art, science, education, and counseling are not nature-connected, whole life organic **when they easily could be.** The addition of Organic Psychology to them is the best remedy for "normal," nature-disconnection PTSD. Time and space to Grok nature have always been there for us since our conception. In fact, the latter can be seen as our most reliable Grokking relationship.

Keep in mind that, a wide range of studies[9], show **a wide range of things go better with Nature.**

Because the 54-sense, self-evident, nature-connecting process in this book emanates from genuine sensory contact with natural areas **it is organic**. It empowers any of us to help reverse today's local and global calamities by organically fulfilling our 54 natural senses in natural areas and their wisdom. This increases their and our well-being to the benefit of all.

Get Real

The natural world, you and I suffer "**Earth Misery**," the degrees of mental illness, disorders, and discontents we create when the story that indoctrinates us, **disconnects us from and violates** the self-correcting life of nature and its beauty in and around us.

You and I recover from Earth Misery as well as improve our relationships when **the story that socializes us re-connects us to nature's Revolutionary Wisdom**. This re-connection organically blends our 54-senses with the natural world's self-correcting ways, a process that we call the art and science of ***Organic Psychology***. This tool enables us to enjoy the healing wonder of Nature's rapture, backyard or backcountry and free of charge.

Be prepared. Don't be vulnerable to Industrial Society and its ploys without letting Organic Psychology help you solve the problems we all face. As the evidence-based essence of any relationship, **the self-correcting power of** Organic Psychology speaks for itself[10].

Really Get Real

Get out of that runaway car in the Core Metaphor described earlier. Although you will receive instructions later on how to improve your connections with a natural area, start this critical process now. **From this point on, only read this book while you are in contact with an attractive natural area or thing.** This makes it an expedition. As you read it, after each paragraph or point the book makes, discover if you can sense or identify something or things

---

[9] http://www.ecopsych.com/2004ecoheal.html

[10] http://www.ecopsych.com/survey.html

about the natural area that is attractive to you. **Do this now.** See if you can sense that finding attractive connection triggers good feelings. This is not an accident. Instead it's their unadulterated origin. A weed, insect, tree, pet or the sky, clouds, and stars are beautiful to connect with when a backyard, park or another natural area not available. Even a picture will do as long as you keep in mind that it is an image, not the real thing.

PART ONE

Our Universal Foundations: Organic Anchors to the Whole of Life

"This nature connecting activity helped me become aware of my attraction to the crescent moon as it hung over two hills near my home. Soon, its mellow glow, framed by peaks and trees, embraced me in a wordless, ancient primordial scene. Timeless power, peace, and unity swept me into awe. I felt in balance with all of reality. I was simply "BEING." No tension, no pressing goal, just truly belonging to the global community. This natural energy captured my stress-laden pulse and seduced it to the rhythms of Earth. The sleeping disorder I have battled all my adult life dissolved in it. For the first time in decades, I gently fell asleep after dark and arose shortly after dawn. I celebrated the breakthrough, and I thanked Earth. I thanked the activity, too, for it enables me to reconnect whenever I choose."

Overview

When not corrected by applying Organic Psychology, the nature-exploitive stories of Industrial Society injuriously disconnect our congress of self-correcting, 54 natural senses from the life of Nature. This painful loss of whole-life satisfaction, that is naturally obtained from gratifying contact with authentic natural areas, desensitizes our senses.

To substitute for being severed from the natural joy, unity, and wisdom of natural areas, our senses are socialized to gain fulfillment or tranquilization from nature-disconnected relationships and attachments to people, substances, and things. Our senses can obtain supportive emotional replacements for their loss of authentic nature-connected gratification **from substitutes like the destructive and senseless torture or killing of life in the form of wildlife, pets, species along with the life of Earth.** This temporarily reduces or releases our hurt. However, it unleashes the same trespasses that earlier injured these senses in us, even though we vow never to inflict that misery on anybody. Our sense of reason sees how stupid this nature-disconnection phenomenon is, but we can't give up our addictive rewards from it. This shames our humanity, and that furthers our hurt and Earth Misery. Applying the art and science of Organic Psychology, and helping others do the same, is a practical remedy for this destructive phenomenon.

Our madness is exemplified locally by the death of Joey and globally by the extinction of species. What similar inhumane circumstances have you experienced or witnessed with people, places, and things? As you continue with this expedition book think about how you would or could apply Organic Psychology to stop this insanity while restoring yourself, others and Nature, to the benefit of all.

Organic Anchor A: The Art and Science of the Senses

Our senses and sensations are undeniable facts of life. Too often we learn to omit their universal life wisdom.

As exemplified by the sensation of thirst, homeostasis in scientific circles is explained, on cellular and molecular levels, by **sensors (senses and their sensations)** in an organism, massive (Earth) or small (nanobe), being genetically created receptors that naturally detect stimuli.

> ***When the information that our senses register is out of balance, they become the primary homeostatic driving force for change that promotes life in balance.*** Their detection process is a fundamental source that functions on mechanical, thermal and chemical levels as it supports the survival of life.

> When our senses are not blocked or adulterated by nature-disconnecting stories, **they can be depended on as undeniable, self-evident, recovery and balancing tools.**

We suffer our problems and hurt because our education denies this scientific truth: As part of the life dance of Nature and Earth, moment-by-moment, we have fifty-four natural senses that are attached to all that has preceded us and all that follows us. Our senses that have been wounded by abusive relationships remain wounded. We feel and act out that pain until we learn how to create space for these senses to happily reattach to the healing wisdom of their origins in the life of Nature's self-correcting balance and beauty, backyard or backcountry.

> **Through your 54 natural senses, in a natural area, your life is as sensed, loved and balanced as part of the life of Earth as is the life of your fingers, heart, and toes devoted to supporting your personal life.**

Consider this quote from *Educating, Counseling and Healing with Nature*[11]:

> "My lifelong communion with trees allows me to know them without sight or language. The beautiful elucidation of fifty-four senses, below, has given me a gorgeous language with which to tell this love story, one I have struggled to share my whole life. I don't want ever to sound mysterious, or otherworldly. For me, this communication has just been a fact of life. But how to explain it to others? I still feel there is an element of my understanding that is nameless, and so loving it needs no words. But to have a natural sense, new brain language, to describe this experience in a way other's can understand, is lovely, just lovely."

Our senses are the parts of us that register life in and around us. We build our thoughts, feelings, and relationships on what they convey. When they are limited, warped or injured so are our lives and happiness.

When you discover a sense or sensation in this natural area/thing you are connected with as you read, find its name and number on the list of senses (See Appendix A: Our Fifty-Four Natural Senses and Sensitivities) and identify the senses. **Identifying the senses is the most reliable way** to accurately connect your story way of knowing with the non-story, 54-sense, way that Nature works in and around you.

---

[11] http://ecopsych.com/ksanity.html

Source of the 54-senses

Between the years of 1961-1978, researcher Guy Murchie made an exhaustive inquiry. He painstakingly scrutinized scientific studies about natural senses, studies that appeared in many hundreds of books and periodicals during those 17 years. The visionary architect Buckminster Fuller was quoted as saying Murchie's book, *The Seven Mysteries of Life* contains: "All the most important information about everything humanity needs to know!"

Murchie maintained that many of the boundaries in normal science are arbitrary; between planet and moon, between plant and animal and between life and non-life. He often makes it very clear when his examples are grounded in empirically verified science. Consider your sense of respiration and your hunger for air that you can bring into your awareness by holding your breath.

> **"With each breath, you take into your body 10 sextillion atoms, and - owing to the wind's ceaseless circulation - over a year's time you have intimate relations with oxygen molecules exhaled by every person alive, as well as by everyone who ever lived."**
> **- Guy Murchie**

In 1986, after my National Audubon Society International Symposium *Is the Earth a Living Organism?* Murchie told me that scientific methodology and research had identified over eighty different biological senses/sensitivities[12] which pervade the natural world and us. He said he additionally verified this with scientists at the Harvard Biological Laboratories. All these senses, he said, he lumped together as 31 senses for literary convenience in his book *The Seven Mysteries of Life* published by Houghton Mifflin in 1978. You can use the index of this book to find advanced answers to significant questions.

Murchie's dedicated efforts deserve our applause, thanks, and trust. He learned to learn most of what he knew from travel and connections with places all over the world.

From Murchie's original collection, I identified 53 (now 54) natural senses, "webstrings"[13] (strands of the web of life) that my students and I had experienced during my 26 years living and teaching on education expeditions outdoors year-round. **These 54-senses are undeniable truths because we can consciously experience them;** they are self-evident. They are our genetic properties that initially came together as part of creation and the stars. For this reason, I warranty

---

[12] http://jhupressblog.com/2012/02/01/how-many-senses-do-we-have/

[13] http://ecopsych.com/webstring.html

the reality of their existence[14], and I have listed them in Appendix A: Our Fifty-Four Natural Senses and Sensitivities. Each sense is trustable in any given moment, and in concert with all of the others. The 54 include the senses of reason, consciousness, and literacy. Also, we have many additional senses that we react to without being aware of them.

Every natural sense is an alive and distinct sensation, an inherent natural attraction love[3] that is conscious of what it is attracted to. Each is a strand in the web of life, a genetically rooted relationship that enables everything in the world to breathe together in balance, including humanity, through universal natural attraction communications, guidance, and motivations. **The life of our 54-senses is as much an actual part of any time/space moment of the life of our Planet and Universe as is anything else.**

Each sense is a consciousness of our love for our planet mother, Earth, and it is a love that is shared by all. Conscious or sub-conscious, each sense is a unique felt-sense actualization of the love essence that Eric Fromm called "biophilia," the love/affinity/attraction to all that is alive[15] including the Universe and its Unified Field that Albert Einstein postulated and that was recently confirmed.

The experience of making conscious sensory contact with the Unified Field essence in a natural area **blends our senses with the area's peace and power**. We heal by becoming aware of self-correcting attractions there that recycle, transform and peacefully unite our disruptive differences.

> **"My life was ecstasy. In youth, before I lost any of my senses, I can remember that I was all alive, and inhabited my body with inexpressible satisfaction; both its weariness and its refreshment were sweet to me. The Earth was the most glorious musical instrument, and I was audience to its strains. I can remember how I was astonished."**
> **- Henry David Thoreau 1851**

Consciousness is but one of 54 natural attraction senses that we share with Nature and that register it. Because the natural world geologically preceded our recent appearance, **what we sense in a natural area, is what in us is doing the sensing[16].** This is a scientific actualization of Panpsychism via self-evidence experience.

---

14 http://ecopsych.com/journalwarranty.html

15 http://ecopsych.com/journalaliveness.html

16 http://ecopsych.com/54rwbook.html

The story world of our Ego, and in general, forgets that the Ego is wearing nature "glasses" that scientifically consist of 54 self-evident attraction sensitivities. Without them, it could not sense anything about the balanced and self-correcting dance of the natural world, in and around us.

Sensory-deficient glasses that omit from our relationships and Ego the existence of homeostatic natural intelligence and consciousness of any of these 54 sensitivity groups produces our unbalanced ways and disorders. We correctly call such stories senseless or non-sense.

If they could label things or speak, the members of the Unified Field's web of life might identify their more than 54 natural attractions, since or before the Big Bang, as their instinct, attraction or intention to diversify in the attractive, loving ways of the universe, moment by moment.

Our 54 natural senses, not our stories alone, are our way of registering the Unified Field[17] as well as being in organic communication with ourselves and others including the members of the Web of Life.

The 54 are a multisensory, non-story intelligence that flourishes now as it did before the arrival of humanity. (For further information see the balanced sensory wisdom and community relationships in action with the Slime Mold and its food[18] also partially described here further on. Learn how and why Nature works cooperatively, even mathematically, in and around us, a process that our prejudicial stories[19] describe as competition and survival by conquest.)

Objective science produces our amazing technologies and contemporary life by only using eight natural senses while society tells us that we have just 5 senses. **Both omit our 45 other natural attraction senses** because they are "subjective" meaning that they can't be measured. **This makes Objective Science only 15 percent whole life accurate.**

A 54th sense, "natural attraction"[20] has recently been added. All 54 of the senses are the diversification of the original life attraction/intention, the homeostatic, Grand Unified Field of Higgs and Einstein[31] (Appendix A: Our Fifty-Four Natural Senses and Sensitivities) that interconnects the life of the Universe. It makes our 54-senses expressions of the Higgs Boson or Unified Field that Einstein predicted.

---

17 http://ecopsych.com/earthstories101.html

18 http://ecopsych.com/journalslimemold.html

19 http://ecopsych.com/prejudicebigotry.html

20 http://ecopsych.com/journalproof.html

As a quiet visit to a natural area demonstrates, whenever our natural attractions in nature genuinely connect our psyche with the web of life, **it energizes and restores our natural senses and their self-correcting ways.** They, in turn, transform our disorders into unified, healthy relationships. The latter contains the balance, cooperation and unconditional attraction/love that the life of nature's spirit shares with us to sustain life in peace. This is universal intelligence because I repeat, whatever any of our 54-senses find attractive in a natural area is that thing in, us, and as us, that is doing the finding in order to be whole. **That is the way the wisdom of the life of Nature and Earth works everywhere to sustain its beauty, purity, and balance.**

> Our leaders seldom teach us that, scientifically, Natural Attraction is the essence of the *Unified Field* of our *Big Bang Universe* as well as the essence of *life, love,* and *unity.* The six are fundamentally identical and interchangeable synonyms. This suggests that "God" is the label that some of us give to **the love of all things for each other everywhere**.

There are, of course, many more than 54 additional natural sensitivities found in the life of nature that humans do not naturally need to register for survival. Ultraviolet light and high-frequency sounds are prime examples.

Most misunderstood are the naturally attractive contributions of the discomforting senses like pain, distress, and fear. We often forget that we might severely burn our hand on a hot stove if the pain did not lovingly signal and motivate us to immediately find some other more attractive place to place our hand. Each of these senses (#25-#27) **serves as an attractive and welcome intelligence[21],** a motivating signal from nature to find additional natural attractions to support our lives (See Chapter 13 in the book *Reconnecting With Nature,* by Michael J. Cohen).

Although Ames, Gesell, Pearce, Rivlin, Gravelle, Samuels, Sheppard, Sheldrake, Spelke, LePoncin, Wynn and many scores of other researchers have, since Murchie, further validated our multisensory nature, the full significance of it has yet to be recognized by contemporary society. Our prejudicial addiction to our nature-separated lives and thinking keeps natural attraction senses and their value hidden from our immediate awareness because they are inconvenient. For this reason, they and we are frustrated by great dissatisfactions and very challenging problems that our non-sense is missing enough active sense(s) to solve.

Our economy fuels itself by making and keeping the life of our 54 senses[13] to be nature-separated discontents. It further irritates them through advertising and then sells us products that satisfy our irritation. However, when unadulterated, or temporarily revived in a natural area, our natural attractions are an essence of nature in action. **Each of them attracts our**

---

[21] http://www.nature.com/news/how-brainless-slime-molds-redefine-intelligence-1.11811

consciousness to the whole life of the natural world and its self-correcting ways, and this includes the natural systems in ourselves and other people.

Each natural attraction sense is an intelligence that helps us sense, feel and love the life of Planet Earth as our other body and mother.

As our good experiences in natural areas demonstrate, our natural senses when connected to nature's homeostatic intelligence, produce fulfillments. They are sensory satisfactions and happiness that reduce stress and its related disorders. Also, the natural connection's "side effects" increase social and environmental well-being rather than deteriorating it.

Any individual who invented a pill that produced the, above, life-unifying results would be a billionaire. However, the pill can't be created. There is no identified substitute for the life of Nature's Dance of the eons, *in, around and as us.*

Whenever our society encourages our new brain story to conquer the life of nature and the natural, we learn to conquer and subdue our 54 natural senses and their expression of our genetic makeup. When excessive, this loss automatically produces disorders.

Our abstracting, nature disconnected, objective sense of reason[22], exalts the few senses that our materialistic stories use to conquer and omit our 47 other natural senses and the life of Nature.

We exploit and demean the remaining 47 natural senses as "subjective," not-scientific. However, their 4-leg, sensing/feeling[22] ways **tell us about how the life of the natural world works its perfection and enables us to participate in the process.** Our absurd disconnection from them makes us only 15 percent whole life intelligent. It explains why we can identify but seldom solve problems in a balanced way.

Overwhelmed and numbed, our 54-senses are a "vast" missing part of a responsible story about the life of Earth, ourselves and community, about how and when to act where._Without the aliveness of all our senses registering in consciousness, we are "half-vast."

As Carl Jung and others have noted, our abstract thinking is no more reasonable or discriminating, logical and consistent than are our feelings.

Our challenge is to recognize that the excessively nature-separated parts of ourselves and our culture are unreasonable.

---

[22] http://ecopsych.com/nineleg.html

We desperately need to think with nature's wise ability to maintain and restore life, without producing our problems. Applying that 54-sense homeostatic wisdom prevents and stops our society's destructive actions against ourselves others and the environment.

> "So, I asked the expert 'What is Gravity?' He explained it as a physical force that Isaac Newton discovered. He disbelieved when I told him gravity was a sense, so I asked him to pay no attention to his sense of Gravity (sense #12), and he soon ended up lying flat on the ground. You try it with yourself or others. We are in trouble because we also learn to ignore the truths and sensibility of our 45 other senses, too. This new frontier of mine[23] is so hidden that it is seldom even recognized as a frontier."

The absence of more than 45 sensory ways of knowing from the life of our conscious thinking is the mother of our collective madness, of our runaway wars, pollution, dysfunction, disease, mental illness, apathy, abusiveness, and violence. They are seldom found in nature. Without experiencing these unifying attraction senses, our consciousness abandons our natural sensory "inner child," and the inner child in other people and species. It hurts and disintegrates the creative sensory passions[5] that bring about community, balance and positive change peacefully.

Our 54 natural senses, in their congress of homeostasis, are our Rosetta Stone of the Universe. They enable us to register the same information in different languages, sensations, and senses, a key to modern understanding, unification and conflict resolution. It allows our eyes, which can see the world but can't see themselves, to see themselves via 54 "invisible" senses.

I offer the list of 54 natural senses in Appendix A: Our Fifty-Four Natural Senses and Sensitivities with this important reminder: Each sense is a distinct, alive, sensory attraction that in nature *has no name for itself* because nature cannot use stories, names or labels. Humanity is the only known species that have this gift.

Each sense can energize the life of many natural parts of us when we use it to connect with the natural world in the environment and people. That touchy-feely, hands-on, 54-sense connecting experience, not the list of senses, catalyzes personal wisdom, growth, and balance.

The 54-senses list only provides information in story language. It places senses on our screen of consciousness and guides our senses of reason and language, through our, story way of knowing. However, without 54-sense passion (apathy), our senses of consciousness, reason, and language are ineffective when it comes to disengaging our common, destructive bonds so we may enjoy responsible behavior, growth, and change.

---

[23] http://ecopsych.com/think3genius.html

For example, even though cigarette labels and research stories show cigarettes to be harmful, many people continue to smoke them. This is because our senses of reason and language are only 4% of our total innate means to know and love nature's life and each other. Our remaining 52 sense groups complete the process. Without them awake and well in our consciousness, we experience apathy and hurt, we don't participate, and our smoking problems continue.

Organic Psychology, nature centered thinking uses the list of senses, below, **in conjunction with visiting natural areas and creating safe moments and space for our indoor conditioning to learn and relate through 54 natural senses in us that we may awaken in nature.** To do this is reasonable, for after we experience a natural sensory attraction, knowing and speaking its right name places that *senses and sensations* in our new brain consciousness. There we can think with it and be motivated by its wisdom. This process non-verbally connects, rejuvenates and educates us. It extends us to safely reach into the natural world to more fully sense and make sense of our lives and all of life. It works because once we experience the balanced life of nature's restorative process and wisdom, we own it. And we never fully return to our former way of knowing.

At birth, some or all our 54 natural attraction senses begin registering self-evident information about the world[24]. When they register flawed information[25] about Nature accompanied by rewarding love/survival, they can, for their lifetime, attach or addict to that short circuit and its harmful consequences.

Educating, counseling and healing with Nature (ECHN) enable us to beneficially reconnect our misled sense(s), including our sense of reason, with the natural world at any point in our lives and reasonably transform our Earth Misery short circuit[5] into the ways and means to deal with it.

The 54-sense list (See Appendix A: Our Fifty-Four Natural Senses and Sensitivities) explains how, sense by sense, nature connects with itself in us, through us and to people and places around us. It shows that we can consciously engage in this process. It validates Dr. David Viscott's proposal that feelings are the truth, that we don't live in the real world when we ignore what we are feeling. Our nature-separated lives disengage and de-energize these senses. Applying the organic psychology of the Natural Systems Thinking Process allows nature, the mother of these senses and feelings, to nurture and strengthen them, to rejuvenate them to normal. The process gives them enough energy to appear on our nature desensitized screen of consciousness and green our thinking.

---

[24] http://phys.org/news192693376.html

[25] http://www.webmd.com/parenting/baby/news/20110526/babies-think-therefore?page=2

"Feelings are a bodily thing, and respecting them is called, and is kindness." - A. S. Byatt

By not incorporating our 54 natural attraction senses in his deliberations, Albert Einstein was limited and unable to "prove" or demonstrate his Grand Unified Field Theory with Physics and mathematical equations. He seemed unaware that his heroic attraction, self-defense and attempts to do this was the "Grand Unified Field" expressing itself in and through him, moment by moment.

Our challenge, as was Einstein's, is to let our senses help us act based on their self-evident, unifying properties. This enables us to contradict, limited or false stories along with their detrimental results."

Individuals trained in Organic Psychology enable the world to build peaceful, green, relationships and economies because our natural senses have unifying powers. If you want to register best what these powers are, recognize this: if you are attracted to reading right to the end of this sentence, this attraction, that you now experience many of these senses in action.

Journaled Project NatureConnect Student Results:

> "The experience I'm recalling is a trip I took with a friend to Mt Lemmon in Tucson. After a half mile hike back into the forest at the 9,000 feet in elevation level, we found a mammoth rock. It was the size of a house and half buried in the earth. We both laid on our brother rock for about 20 minutes. It was a cool sunny (one cloud) day, and we were surrounded by a forest of massively beautiful pine trees.
>
> The rock was oh so warm and comforting to embrace as the sun shone on my body. The contrast of the warm rock and the cool breeze was so wonderful. Wind from my entire body. A physical therapist would have charged me $150 to do what nature did for zipping.
>
> My senses registered colors, sounds, feelings, aromas, sensations, moods, contracts, textures, sizes, distance. I stopped controlling the world and let mother earth breath for me. I felt the texture of the rock and the sensation of holding up the world on my back. It was weightless and comforting to support. The sounds of the nearby creek and the many bustling creatures were a symphony of natures' voices all welcoming me to stay as long as wanted. The smell of pine was in the air. All senses were on the maximum open channel, and I melted into the moment with ease.
>
> It was only my natural sensory attraction connections to the natural area that provided these rewards. I have never, in all my years of public or higher education been taught anything about what I just experienced."

"Walking along the edge of the coast, I saw three White Egrets and One Great Blue Heron. The water seems to be a clear dark blue at the edge. (#30 sense of physical place) I feel the wind in my hair, (#14 feel touch on the skin) with air with a slight chill to my skin that made my body feel chilly" (#7 sense of temperature)

"The sky was filled with pink multi-dimensional clouds that seemed to radiate out above and all around me (#4 sense of light and moods attached to colors, #16 space/proximity sense, #41sense of form and design). Absolutely incredible!"

"Wow! I said to myself. It seems to be radiating right toward me."(sense #39 language and articulation sense, #43 sense of consciousness, and #35 sense of self including friendship, companionship, and power)

"I found myself running through the crisp morning air (sense of temperature) to get a clear view." (#42 sense of reason, #18 sense of motion and space and #5 awareness of one's own visibility)

"I sang out my thanks (#35 sense of emotional support, belonging, support, trust, thankfulness). As I twirled, arms outstretched, the sky turned from pink to gold (#18 sense of motion, #29 play, pleasure, laughter, place, #17 coriolus sense or awareness of effects of the rotation of the Earth). The rising light danced off the expanding shapes." (#4 sense of sight, color, #41 distance, and design)

"I called my truth aloud, 'Thank you dear clouds for helping us to feel our connection to all things and to all living beings!!!!'" (#40 sense of appreciation, humility, and ethics)

"I felt as if I was being watched over and supported." (#44, #43 sense of intuition, deduction, sense of mind and consciousness)

"All of my personal senses and sensations took place in, were part of and contributed to the now, the present time-space moments that started with the Big Bang 13.8 billion years ago. This means that what I love to sense, think and act now will be part of and influence the next moment of the life of Earth. This feels good. It helps me realize that there is a level, supportive playing field in Nature and that my life contributes and has value and can help shape it in a good way.

Since we were children my older sister, and I have always argued, but after I connected with the clouds, I felt a special sense of self-recognition and appreciation. I recognized that feeling was what I was missing from my sister and that I did not need to get it from her, I could get it from the clouds and other nature connections, and I continued to do so (#43 sense of consciousness, #42 reason, #39 language/stories). Things are better now. I no longer argue with her, and we respect and love each other more." (sense #54 whole life attraction).

A Self-Evident Fact: We have at least 54 natural senses.
A Supportive Observation: *Show me a person who is convinced that we only have five senses and I'll show you a person who never learned to count to six.*

Organic Anchor B: The Tree of Life

After fifteen years of outdoor and expedition education, the core of this book began in 1965 at the bottom of the Grand Canyon wilderness in Arizona. There Mike Cohen put into words[26] what he was feeling about nature. "It was self-evident. I could not deny it. I experienced the life of Earth as an organism that has the will and knowledge to help itself survive. There was nothing I found that I did as a living organism that it did not do, **except that I could put things into words and create and tell stories. I could then think and feel about what they said and act accordingly while the life of Earth and Nature was oblivious to them.**"

Today, 57 years later, along with Stacey Mallory and many others at Project NatureConnect, through his evidence-based words here Cohen scientifically assures you that the life of any tree emulates the 13.8 billion-year old life of the Universe, a life that we call Nature or Bio. **The tree looks like Nature's aliveness because it is it.** Every tree is a living hologram of the universal "Tree of Life." Their essence is identical. They all sprout and grow from the ancient Big Bang **seed of aliveness and its attraction to produce its time and space sequence,** moment by moment over the eons to this moment. For example, the double helix of DNA is the same form as the path of the Earth around the sun for a year as well as the whorl of the branches of some trees around its trunk. (See Eco-Art Therapy **Appendix E: Tree-ness**). Scientifically, this blends with the Tree of Life that is the beginning of life in most spiritualties and religions. (Hint: by the end of this book, scientifically you will be able to relate to that tree in natural areas as the "Tree of Love.")

In evidence-based reality, a tree is the life of Nature manifesting itself as that tree living its life, a phenomenon that is best called "treeness" or "treeing." Equally valid is that you, I and everything else is the life of Nature manifesting itself as the life of us being who or what we are. We all sprout and grow from that same ancient seed of aliveness and its growth sequence of the eons that also produces treeing. Trees could be called "tree people." In some Native American languages, there are no nouns; everything is called its own, special "ing."

---

[26] http://www.ecopsych.com/livingplanetearthkey.html

Recognizing the fact, above, for yourself makes life more understandable, peaceful and united. Just visit a natural area and note that you have walked into and are visiting your natural self naturally loving to live and successfully doing so without your ability to tell stories.

Observe that everything in the area, including you, is starting now or has had a starting point. Everything is on the same moving train of the Universe and its sequence; everything was created and started by it at various locations of it along the way. This is because they/we are all the same original Big Bang process of the life of ancient Nature a loveness or loving to grow and to be itself into this moment and the next.

Observe that this natural area space, including you, had an alive beginning when you entered it. Now, in this moment, it continues to operate and, like a seed, grow from the aliveness of that beginning of this time and space a few minutes ago. You are observing, sensing, feeling and being the Tree of Life doing the same as it has since the birth of time and space Big Bang origin. You are it making itself at this moment, including you, and it is self-evident that you are experiencing it and its joy. In the now of this natural area, you are yourself meeting yourself as a new, seamlessly continuing, part of the original Big Bang seed and Tree of Life. So is any seed.

Get to know yourself better. Intellectually, emotionally and physically, when possible, thank or embrace anything and everything in the natural area that is safe and attractive to you. Enjoy how balanced, beautiful and sane it, and you are. Validate that you are not a sinner, wrong or bad for loving yourself as your natural life.

Our society's past century of painstaking scientific study has determined that the life of Nature started with a single, life-loving, Big Bang "fertilization" of its "seed" of the Universe. The unity of the number symbolizes the moment "One" being itself rather than not being at all, "Zero." Oneness or One-ing was attracted *to be* rather than *not to be*. That was not a question, Mr. Shakespeare. :) There is no question "To be or not to be?" because of whatever or whoever is already being if they can ask the question.

Throughout the eons, the oneness seed of the ages has loved to germinate, grow and reproduce. It continues this process. It is Nature's and our aliveness at this moment and the next.

Restated, in a natural area, what we call Nature or Planet Earth is the universal "Tree of Life" loving to be alive and continuing to live. This is also the profound feeling that we call our sense of self, our love of life and our life. It is our powerful, self-preservation, desire to be, to survive. It is the way we born and are socialized to experience and relate to the essence of the Web of Life.

From the life energy of an atomic particle to the aliveness of our planet and beyond. All things in the plant, animal, mineral, and energy kingdoms are attractively and cooperatively engaged in continuing the life of Nature's original unified seed-to-tree process and its sequence of its self-organizing ways, right into this moment and the next.

From its unique love to live and live to love origin 13.8 billion years ago, the seed's web of life today includes the aliveness of our love to sense, think and feel in a unified way, "to be," and experience the happy wisdom of our natural oneness. Scientifically, that singularity was Einstein's Unified Field source of life, love and everything else leading Carl Sagan to say "If you wish to make a pie from scratch, you first have to invent the Universe."

When our information, stories, and relationships scientifically **support** the life of Earth, as above, we benefit from our congruence with the organic balance, purity, and beauty of Nature's self-correcting way. Its unified origin and integrity speak through and to us because it is us.

When our stories **don't support** the life of Earth, they disconnect us from its love and perfection. We suffer because we are rewarded or punished to replace this disconnection loss by accepting or seeking artificial substitutes for it. When this happens, **we emotionally attach or addict to our flawed or limited imitations of the life of Earth along with their detrimental side effects.** That is the source of most of the problems we face.

Nothing escapes the energy of the seed-to-tree sequence including knowledge so we can't fully understand advanced information if we do not understand its earlier components. For example, new numbers have little meaning if we don't first correctly understand prime numbers. For this reason, much of education, science, and history consists of evidence-based discovering how the past worked. That's what helps make it useful in the present and predictable for the future.

In retrospect, whenever we visit a natural area and recognize it as the eons of Nature's "Tree of Life" continuing to build the time and space of its life in the moment, we permit our 54 natural senses[27] to register and consciously blend there with the original organic aliveness and unity of Nature's life-sequence at work. The wonder or rapture we find there is a grandeur that many folks at times exalt. Its enchantment relaxes them. However, they seldom can scientifically explain this phenomenon. It mostly expresses itself as a wonderful personal spiritual or emotional experience that they revere. They'd love to share or give to others but can't. It is personal, and they don't have the organic tools they need to help others experience it. They are missing the Revolutionary Wisdom art and science that enables them to share it by experiencing it together in a natural area. Its Organic Psychology in action is the missing "technology of behavior" tool that in 1971 B.F. Skinner called for in *Beyond Freedom and Dignity.*

---

[27] http://www.ecopsych.com/insight53senses.html

By applying the revolutionary wisdom in this book, you can **GROK** (sense and validate your oneness with something) and benefit from your origins as well as share them. Accompanied by the list of our 54 nature-congruent senses in Appendix A: Our Fifty-Four Natural Senses and Sensitivities, go to a natural area and gain consent from the life of Nature to apply that list there and identify the 54-senses in that natural space and time place. Groking helps you sense, think, feel and benefit in the moment from the totality of the Tree of Life. All your senses consciously unite with their origins at that moment. That embrace provides you with their revolutionary wisdom and happiness.

Your newly gained Revolutionary Wisdom story enables your 54-senses to experience Nature's unity at will[27]. They fully register the magnificence of being the Tree of Life's love for its attractive, self-correcting, speechless and peaceful sequence of birth, growth, interaction, diversity, balance, connection, cooperation, communication, reproduction, aliveness, transformation and recycling of each part of the Web of Life, all this happens without the Tree producing death, garbage or abuse. (Does this leave you speechless? Does it help explain why Nature can function even though it is speechless?)

The ***Tree of Life*** is a natural being, a blueprint, created to ensure our survival. Bring into your mind's awareness a tree you have gone by or one that is outside your window. Envision all renditions of the tree, from birth to the one you see now, including its future "death," are all inside one, now, space and time in the Unified Field. The Tree of Life never really appears, dies, or moves on its own in the moment, it is only our long-term view of reality which gives the tree the appearance it is evolving. (See Eco-Art Therapy **Appendix E: Tree-ness**)

Our treeness experience can help us recognize how much we limit our personal and global information and well-being when our excessive disconnections from Nature reduce us to mainly knowing the natural world as "wilderness," "uncivilized," or a "resource."

You can achieve certification or a degree for applying the Tree of Life and Nature's wisdom as a much needed art and science. This is because our most challenging personal and environmental problems result from the difference between how our nature-disconnected selves unreasonably socialize what our life senses and feels, and how the life of Nature wisely works to sustain it's/our perfection.

Our lives are too often attached or subject to people whose stories and experiences excessively disconnect them from and make them abuse the Tree of Life essence of Nature/Earth, in and around us. We all suffer accordingly because most of us are socialized to **spend over 99 percent of our life out of tune** with the way the balanced and beautiful life eons of nature work now. We are troubled because our recreation is not spent in re-creation, through reconnecting with the natural world. Too often the side effects of this omission have adverse impacts that make it "wreck-creation." We teach ourselves not to love natural re-creation and instead love costly recreation. For example: When we don't choose to take a peaceful, cost-free walk in a natural area and to play golf instead. Golf has changed natural areas into 3,507 square miles of golf courses along with the chemicals and water that maintain them. That is three times the size of Rhode Island.

To help understand the significance of this book/course and Project NatureConnect to be aware of our troubles from Earth Misery Day[28]. Then consider that due to our excessive disconnection from Nature our dreams are mostly about gaining support due to the hurt and stress that accompanies our excessive disconnection from Nature. As many PNC folks will tell you, PNC helps us achieve that support. This is because our attraction to daydream mind-wandering accounts for almost 50 percent of our waking hours involved with additional disconnections the life of Nature rather than to benefits from reconnecting with it. Our sleeping time dreams are similarly engaged with the effects of our excessive disconnection. This book helps the world solve its runaway personal, social and environmental problems by making these waking and sleeping hours, along with video game time etc, be advantageously connected to the life of Earth's revolutionary wisdom for 75 percent of our daily life, instead of, as at present, our lives are 99 percent of our lifetime out of tune with the life of our planet.

> **"For the individual, mind wandering offers the possibility of very real, personal reward, some immediate, some more distant. These reward include self-awareness, creative incubation, improvisation and evaluation, memory consolidation, autobiographical planning, goal driven thought, future planning, retrieval of deeply personal memories, reflective consideration of the meaning of events and experiences, simulating the perspective of another person, evaluating the implications of self and others' emotional reactions, moral reasoning, and reflective compassion."** (Singer)

Our nature-disconnected stories compel us to separate our mind, body, and spirit from the wellness of nature's life, in and around us, especially during our formative years. This attaches the essence of our being to many questionable artificialities, sold to us for excessive profit and economics, instead of the Tree of Life. All the while, our leaders are fully aware that our Planet has been resource bankrupt for the past half century and, being identical, so is our life.

We have turned the health and peace of our planet's aliveness into an insane asylum run by its inmates. We have learned to build a madhouse society that celebrates its ability to clear-cut the Tree of Life and produces addictive commodities and tranquilizers to subdue the pain this severance produces temporarily. All the while, the revolutionary wisdom remedy for the stress, sickness, and chaos that we create is available, free, in our local park, pet or flowerpot. That 54-

---

[28] http://www.ecopsych.com/zombie2.html

sense wisdom enables us to remedy our pain and disorders and reduce our overuse of natural resources. This book empowers anyone to apply and enjoys this process.

We desperately need **heart-centered individuals** who want to improve their lives by learning to use and teach this book's organic methods and materials. In their spare time, alone they can redress our violation of the life of Earth's integrity, reverse its catastrophic effects and enable us to teach others to do the same. Like taking a drink of water when we are thirsty, each new connection we learn to make with the life of Nature/Earth improves its and our balance, wellness, and happiness.

This Revolutionary wisdom expedition provides us with the purifying and balancing "organic technology of behavior" tools we need **that are being withheld from us**[6]. It helps anybody increase personal, social and global well-being by learning how to create moments in natural areas that let Earth teach us what we need to know and that we have been taught to forget. Our fifty-four natural senses help us hug the Tree of Life and invite the self-correcting ways of the natural world to revive our missing natural wisdom. **Anything less has proven not to be enough since 1974.**

We suffer our problems and unhappiness because we are convinced to deny this scientific fact: as part of the time-space life of Nature and Earth's dance, moment-by-moment our fifty-four natural senses attach us to all that has preceded us and all that follows us. **Senses that are wounded by abusive or unfair relationships remain wounded. We feel and act accordingly until we create space for these senses, with joy, to reinstate their origins in the dance of Nature/Earth's self-correcting purity, balance, and beauty, in, around and as our neighbors and us.** Revolutionary Wisdom helps you happily know the "who, what, where, how, when and why" that empowers you to accomplish this critical necessity. It works because although addiction is an unreasonable, overriding love, the fundamental nature of love is the essential love of nature and its self-correcting powers.

The trustable authenticity of this book's science and evidence is encapsulated, reviewed and refereed in an updated *International Journal of Physical and Social Science* June 2017 article "The Scientific Core of all Known Relationships[3]"

> "This nature-connection activity attracted me to lie in the water beneath that sky of reddish and orange colors, with the sun rising more powerfully than I've ever seen it before. And there I was, all alone in the pond, but at the same time feeling held by the outdoor moment. I knew this is where I belonged. I felt a powerful energy being given to me by everything around me. I had a feeling in my throat that made me feel like crying and shouting with joy at the same time; and, a sense of peace, like everything, was telling me 'It's all right, it's all good.'

My sense of isolation, this sense of loneliness, of being abandoned, of having the world against us is a common feeling. But how could a person be lonely when they could feel like part of a pond or see themselves in a lovely flower, in the trees and animals, and in other people? Think of all the lonely people who have no idea that we are so connected to nature.

We are truly not alone but only disconnected from our natural state. I used to be so depressed, and now when I feel it come on me, I think about how connected I have become to all things. I am less and less depressed now. Imagine if everyone could experience this outdoor connection. It would become a different world." -KM

Organic Anchor C: Sense, Feel, and Act Globally

If the life of Earth died, so would your life. With regard to aliveness they are identical.

Get Smart by Getting Real

The essence of education, counseling, and healing is to give people the things they need to grow into contributing and wholesome members of their society. To achieve this, what was necessary in ancient times has little in common with what is required now to live a contemporary, rewarding life in balance to the benefit of all. Today's advanced but limited science, knowledge and technological practices increasingly produce destructive and unhealthy side effects that we fail to address. It is as if our prejudice against nature has us fighting and winning an undeclared war with the natural world. Earth Misery day has demonstrated, the 1970's slogan "Think Globally, Act Locally" has not worked because it overlooks that our thoughts and acts are socialized to conquer nature, so that is what we have continued to do.

To reasonably maintain and support the aliveness of our personal, social and environmental well-being, the core of what we learn must respond to our modern needs and relationships. This includes loving, rather than excessively conquering and exploiting, the life of the planet we live in. If we do not scientifically strengthen the arts, thinking, and laws that make our relationships connect with the self-correcting ways of Earth, our dilemmas will continue to increase[5].

Think Organically and Act Globally: Sense, Feel and Relate Reasonably

Revolutionary Wisdom enables us to relate to today's disastrous personal and global challenges effectively. It teaches us how to let the life of Nature do what its peace and therapy do best. We learn how to consciously bond with the restorative organic powers of Nature *in authentic natural areas* through all our fifty-four senses, *not just five of them.*

> ***Thinking with only five senses is like having a lobster operate the control tower of Chicago International Airport***
>
> *Revolutionary Wisdom* empowers you to learn, and teach others, to participate in the organic integrity that you share with the life of our planet and society. You can help the essence of Earth's web-of-life increase your personal, social and environmental well-being. In a natural area, it enables your veiled or injured ways of knowing and relating to *Grok* the integrity of Nature's essence. That organic reality replaces the pictures that we purposely omitted on these pages, images that would ordinarily illustrate this book.
>
> In Organic Psychology, ***Grok*** means to validate how you are emotionally embodied in what you are attracted to in a natural area. In the life of Earth's geological, time/space sequence, that thing evolved first. Its wisdom of the eons attracts things to each other in every moment since then, including our fifty-four natural senses and us. Grokking things that are natural has a capital G because it is whole life organic and mutual. Otherwise, you can grok anything including a machine gun.

Anyone can happily, apply the truth of unadulterated sensory scientific evidence to correct the information and emotions that presently mislead us. Reasonably increase peace, sustainability, and justice as you attractively improve your communication with yourself, your friends and the natural world. Help stop the abuse of the life of Nature, ourselves and each other.

We invite you to explore and live a natural area's unadulterated, unconditional love and attraction essence of Albert Einstein's Unified Field of Nature discovery. Use its therapeutic ways to remedy our excessive trespasses against the living integrity of Nature and Earth in and around us. Optionally achieve a degree, certificate or livelihood in this hands-on recovery process.

> [**NOTE: If you want to do this book as a course or degree, with or without credit, contact: nature@interisland.net.** Potential final exam questions appear throughout the text and in PART SEVEN. If you find you can answer them, and they are of value to you, your growth and success in the course are assured. This book/course is designed for individuals who have completed Project NatureConnect's Orientation Course, ECO 500A[29] and its experiences there. If you have not taken that course, **you will find it very**

---

[29] http://www.ecopsych.com/orient

> **helpful to read about it now** in the article *Explore Nature's Wisdom in the Unified Field*[30]**.]**

Most people and disciplines want to stop the madness of our destructive relationships and peacefully increase personal, social and environmental well-being. They are disheartened that our abusive society does not provide them with the ways and means to accomplish this. Uniquely, *Revolutionary Wisdom* provides you with the art of that science.

It makes sense that the life of the Universe/Nature/Earth is attracted or loves to live its balance, beauty, and integrity, in and around us because that's what life is and does. However, us merely knowing this seldom stops our abuse of life because the detrimental, unattractive ways we learn to think and act are too often socialized, habitual or addictive. Our leaders cannot help us stop them because they are us, they have the same habits and addictions.

Revolutionary Wisdom empowers us to scientifically create sensory attraction moments in natural areas, backyard or backcountry, which let the self-correcting life of Universe/Nature/Earth consciously connect our fifty-four senses with the wisdom of Nature's perfection. Our application of this distinct tool enables us to happily transform our abusive attachments into more sensible ways of thinking, feeling and relating as we help others do the same.

> Early readers of Revolutionary Wisdom say: "This book is like putting on an 'intelligent happiness hat' that enables your mind to register forty-seven otherwise silent callings in a natural area. You not only understand them, **but you can also enjoyably think and act with their wisdom of the ages and teach others to do the same.** These callings were a natural part of you at birth, but a government imposed 'indoor socialization hat' that you must wear uncomfortably disconnects you from them.

Natural intelligence and mental health studies with Project NatureConnect include your life experience and prior education in combination with the humanized physics of Albert Einstein and his GreenWave Unified Attraction Field[31] along with newly discovered Nature constructs. With thanks to the pioneering work of Dr. Stacey Mallory[32], **the core reality of our lives as an autobiography can be included or simply be the central focus of factual discussions, academic**

---

[30] http://www.ecopsych.com/Explore Nature Article.docx

[31] http://www.ecopsych.com/journalgut.html

papers and degrees[32]. Revolutionary wisdom helps us find and replace the disturbing elements of that core that were produced by our "indoor socialization hat."

The revolutionary wisdom that you find in natural areas enables you to add hands-on, updated, problem solving, nature-connected processes and professional credentials to your health, loves, relationships, livelihood, and community.

Designed for independent study and members of the GreenWave-54 Grand Jury[6], *Revolutionary Wisdom* is an Organic Psychology expedition book and course for reconnecting with Nature in local natural areas instead of substituting for their integrity. The substitutes too often abusively indoctrinate us to dominate the life of Nature, in and around us and suffer the painful consequences. To remedy this, *Revolutionary Wisdom* enables us to stop our suffering by learning how to create time and space in a natural area that lastingly reinstates us back to health and happiness.

Here again is this book's secret to success. **Only read it while you are consciously in contact with a natural area or natural thing, backyard or backcountry, that most attracts you. Start doing this now if you haven't already.** *The more natural the area, the better.* A weed, insect, tree, pet or the sky and clouds are fine when backyards, parks and other natural places are not available.

During your first natural area contact moment, and each moment after it, you are a seamless, whole life continuum of our scientifically validated space-time Universe[33] and its Unified Field. That contact moment is the *Truth of Now* (TN) **the eon's source, at that moment, of your 54-senses and all other things EXCEPT for stories that *deny or prevent* this phenomenon.** The Universe is united. All things in that instant are attached to everything that preceded it and all things that follow it. As your 54-senses Grok it, you blend with and into its universal knowledge and integrity. This is because since Nature's birth the Unified Field has acted as its own form of internet. It connects all things with the same attractions that formed them. This means that any attractive natural thing or phenomenon you register in a natural area is part of your life expressing itself.

NOTE: Days or weeks may pass as you continue "doing" this book on your schedule and during this period you may forget some key points. For this reason, some points are continually repeated in the text.

---

[32] http://www.ecopsych.com/proposal.html

[33] http://www.ecopsych.com/spacetime.html

Organic Anchor D: The Truth of Now

GROK: How, Why and When

The Scientific Truths of Nature's Wisdom: getting the big things right, an objective search for Mother Earth.

> **Grok the Unified Field sensory wisdom of authentic nature-connected relationships. This is the art of scientifically tapping into the timeless, self-correcting energies of our 54 natural senses to produce personal, social and environmental well-being.**

In any natural area, this book's expedition adventure helps the genius of our **54-senses** *Grok* the life of Earth and happily improve our relationship with self, society, and nature. Learning how to do this so you can help others do it is the core purpose of this guide.

Grok is the *Truth of Now* (TN). It means, in a specific time/space moment, to be fully unified with something, to understand and empathize with it. We do this to the extent that thing becomes part of our sense of self-knowing that we exist in its embodiment while simultaneously we can sense ourselves as it being part of us. In a natural area, this is like intellectually and emotionally **being at one with nature in a felt sense, heart-centered ways in any given moment,** so that next moment includes us being part of it and vice versa, in a good way.

Since Grokking can be a raw, story-less attraction in action, when this occurs in a natural area it is our genetic makeup recognizing and embracing its origins in the Unified Field and celebrating this reconnection by producing feelings of joy. This kind of Grok can equally occur with any of the **54-senses like another person, or yourself** as long as it is organic and not a nature-disconnected story or attachment. The latter do not qualify for a capital G.

Grokking is how things in the speech deficient web-of-life know each other through immediate attraction relationships rather than by our abstract storytelling, that is foreign to them. It is the period when our attraction to wholeness (sense #54, see Appendix A: Our Fifty-Four Natural Senses and Sensitivities) blends with a specific attraction in a natural area. Identifying from the Appendix A: Our Fifty-Four Natural Senses and Sensitivities list **the names of our 54-senses that are activated** is essential to Grok the area. This is because for us to be whole and reasonable (sense #42), we must include our unique story-telling and labeling ability (sense #39).

A Coursework Hint: You Grok! Grokking helps you remember that *you are the one who is beneficially learning to Grok in a natural area and help others do the same to the benefit of all.* If you believe that you can do this then, scientifically speaking, your coursework and degree hypothesis must include you. It would be "**By beneficially learning Organic Psychology through Grokking I can strengthen (your major interests) and teach others to do the same to the benefit of all.**"

Too often our habitual omission of our 54 natural senses desensitizes us to the fact that we are the ones doing the Grokking. This means that in Applied Ecopsychology, you must keep in mind that *you are the one who is doing the applying.*

Also, be aware that a thing in the natural area that is attractive to you, no doubt, geologically appeared in the life of Earth before you. When that thing is attractive to you, it is because it is Grokking you as part of it. That is how and why your 54 natural senses keep you included in and in contact with natural systems. For example, how thirst includes your life in the life of Earth's water cycle.

It also often helps to TN Grok a natural area by finding a natural attraction there and then state why you love it. "I love this flower because I feel its aliveness" Then make the same statement, aloud, about why you love yourself and validate that part of yourself once you find it. "I love myself because I feel my aliveness." You can produce wonderful discoveries through this Grokking activity[34] as expressed in student journals.

> "I felt a strong attraction to an empty snail shell on the beach, and I discovered that I love me because I am imperfect but beautiful. Parts of me have broken away leaving jagged edges and holes. My surface is rough and blotchy. But I have a feeling of mystery and complexity despite my apparent simplicity. I am smooth at my core and I have a secret part that you can't reach unless you are really really small and need a home. But at my center I also have an openness if you look carefully. Wow!!!!!!!!!!!!!!! What manner of sorcery is this!!!!!!!!!?????????????????? I recognize the value of this activity just as I recognize the connection between me and this shell. I surely have a lot in common with it. I feel like I found a gleaming white salty piece of me." -**SC**

* * * * *

> "My attractions to everything in this natural area show me that I like myself because I am the Soul. Pure and true. I am present everywhere. Omnipresent Soul. I connect everywhere with natural attraction and connection. I am as strong as a tree. It is strong as it holds the earth and the earth holds it. I give my love and care to all the earthlings as a tree does." -**RH**

---

34 *http://ww.ecopsych.com/giftearthday1.html*

*****

"I like this Pine because it stands a little apart; it's a great destination, it has a lovely shape and color. I like the curved nature of its branches and its soft needles. And because the tree and I are identical with respect to the natural systems that sustain us, the Pine helps me identify myself. I can truly say I like myself because I stand a little apart; I am a great destination; I have a lovely shape and color; I like my curved branches and soft needles. In metaphor, all this feels right.

Lois hesitated and then read on, Right now I have tears in my eyes. I hardly understand why. I feel as if I can see the Pine now, I feel her essence, her beauty as she stands a little apart. She mirrors something within me. I see that I am transformed when I am outside and connecting to the natural world. I am nurtured, calmed, soothed, held, warmed, excited, stimulated, active, empty and yet full, happy, content and connected."-**LFG**

We suffer because contemporary counseling, education, and healing do not require us to learn how to scientifically Grok the life of Nature moment by moment (**m/m**). This **omission** makes all the information in the world that is now in our possession, m/m deteriorate the integrity and aliveness of our planet and ourselves[5].

***Revolutionary Wisdom*'s expedition enables you to TN Grok Nature at will** because you are required to read it and do its activities while in contact with the authenticity of a natural area, backyard or backcountry. There your fifty-four senses genuinely reconnect, and in time fuse, the healing power and unity of the natural world back into your awareness, life, and society. Most of us have experienced this when watching a sunset or hugging our pet. Grokking in Nature, at will, produces similar self-evident "facts of life" to increase personal, social and environmental well-being.

Build Your Personal Revolutionary Wisdom Safe Deposit Box

In your mind build a secure, empty, sterile storage box or closet whose door only you can open. Each time you discover or Grok a fact of life of Nature/Earth from this book that you know is true for you, lock it in that box so it is not adulterated by falsehoods or other things that may surround you but are not right for you. Open this protected closet when you want to access the TN information and its integrity that you have carefully stored there. It relaxes and restores you because it TN connects you to your original Grok experiences in the natural world that are always alive in you. Opening the closet door energizes them and brings their benefits into your TN consciousness.

Grokking is Significant

The whole world validates **the truth that you are alive and breathing air right now**. Most importantly, you and everybody else trust this "Now" truth.

The *Truth of Now* (TN) is an inborn, self-evident and unifying means of relating and communicating that we naturally hold in common. To our loss, our established ways don't require us to Grok, so we disregard or conquer TN.

By omitting or denying TN and not Grokking natural areas, the world is increasingly unable to unify. In painfully destructive ways we see, sense and feel the life of Earth falling apart in individuals, society and the environment. Without TN we are unable to stop our expanding lack of peace, sanity, and resources while we escalate our extinction of species, climate change, mental illness, obesity and most other disorders. (Grok this now. What does nature "say"?)

Why wouldn't you want to use this book to Grok the source of TN and therapeutically produce personal, social and global unity? (Grok this now. What does nature "say"?)

The authors learned much of the information in this book from direct contact with attractions in natural areas and nature-loving people. For this reason, after most of the paragraphs here you will be able to discover, Grok and validate the truth of the message of the paragraph in the natural area you visiting and connected to. **This practice throughout the book is its essence. It beneficially grounds you in the origins and eons of the authentic life of Nature/Earth and gives you the opportunity to connect directly with its truths. *It is required of all readers* who seek the credit and special advantages that this book as a course provides.** It is critical because it lets the life of Nature "speak" its peace to and through your fifty-four senses including your sense of reason (sense #42). That is why it is reasonable to do this. To be part of a system anything, including yourself, must be in attractive contact with the system. (Grok this now. What does nature "say"?)

During the period that you Grok-experience an attraction that connected with you in a natural area, you are consciously registering the history of the Universe manifesting and expressing its TN space and time as you. You can put your name on it and give it your social security number. That Grok is your unadulterated personal history of you being unified with, in and as the natural world.

- It is it/you consciously yet namelessly producing your TN space and time.
- It is what you might be experiencing if you were born as, or transformed into, that attraction or any other member of the web-of-life.
- It is your natural self, your non-story body, how your pet may feel about you.

You become human by accurately giving the Grok its name in contemporary society. That name is one or more of your TN 54 natural senses, your natural sense of self and self-worth in this time and space. Your ability to label it, correctly or not, is what makes you human, including you calling it a Grok. Your attraction to Grokking is sense #54, your attraction to be part of and experience the whole life of Nature/Earth rather than be isolated from its love. Your wholeness consists of your ability to create nature-accurate stories combined with what the story-challenged life of Nature/Earth in you knows. This technique can help you become more whole (holy):

Write after each paragraph a few thoughts about:

A: The life of YOURSELF (self): what attractive things you found in the paragraph for yourself.

and

B: The life of PLANET EARTH: what value the paragraph contributes to the well-being of individuals, society, and nature.

For example: *Without producing garbage, Nature creates optimums of life, diversity, cooperation, balance, and well-being while, on average, our society detaches over 98 percent of our life from this wisdom.*

Life of Self: My life may be too separated from Nature.

Life of Earth: Our planet has natural wisdom. *(Grok this now. What does nature "say"?)*

When you are sure you have Nature's approval (when to *Grok again is attractive to you),* you can confidently continue to the next paragraph.

Be sure to journal what you discover or what Nature "says" in this Grokking process so that it is available in additional course activities and for your application of Organic Psychology to your individual interests and livelihood. Be sure to put what purity you discovered in your Wisdom Safe Deposit Box, too.

A Grok expresses the greater-than-human life of Earth and Joey. That life matters if we expect to live. By omitting it, we create the problems that we suffer and can't solve. Presently, to our loss, we are socialized to spent 99 percent of our industrial lives omitting it.

Grokking becomes easier as you repeat it because you discover and validate anchors in yourself that you know you can rely on as you continue. For example, the whole world recognizes the truth that you have to be reading these words right now. Most importantly, you and everybody else trust this "Now" truth.

Why wouldn't you want to use this book to link to and Grok the source of TN and therapeutically produce personal, social and global unity? If you can't find a good reason, Grok on. You are experiencing whole-life love in action. *(Grok this now. What does nature "say"?)*

Nature's self-correcting, pure and peaceful ways **are not a miracle, spiritual or supernatural magic.** What the scientific essence of the life of Nature and Earth has, and what we share, enables the natural world to sustain its balance and beauty in and around us. It is sense #54 the sense of homeostatic unity, of instinctive *natural attraction* aliveness and survival. The singular essence-diversity attraction dance mother of all our other senses and the Big Bang in the now moment. *(Grok this now. What does nature "say"?)*

Like learning a new language by visiting the country where it is spoken, this book slowly lets the life of Earth teach you how to be literate in its language and ways. This takes time, and if you do not come to class in a real natural area, you will not succeed no matter how well you do on a test about the facts in this book. The facts are stories and Nature does not speak or understand them[35]. It only relates to how they help us act in balance[36]. *(Grok this now. What does nature "say"?)*

To just read about the scientific art, method and purpose of Organic Psychology, visit its Overview[3], Defense[6], and History[11].

Where is your life with regard to the information, above? How does it affect your past and present, your dreams and future?

A potential final exam question: *Does Grokking a natural area make sense?* (*Who, What, Where, How, When, Why)*

Rationale

Over the eons, Nature/Earth's balance of its optimums of life, diversity, beauty, purity, peace, wellness and cooperation has not shown to produce contemporary humanity's runaway garbage, stress, and abusiveness. **It makes no sense for us to continue to use our flawed processes that cause these troubles.** We must instead establish adequate time and space for our fifty-four natural senses to connect with Nature so we may beneficially think, relate and blend with its

---

[35] http://www.ecopsych.com/nhpbook.htm

[36] http://www.ecopsych.com/eartharticle.html

wisdom. This includes transforming some or our established procedures into more appropriate practices. *(Grok this now. What does nature "say"?)*

> **"My name is Chellis, and I am in recovery from Western Civilization."- Chellis Glendinning**

Organic Anchor E: The Art and Science of Organics and Natural Area Connection

> **"The difference between past, present, and future is an illusion; they all exist simultaneously." - Albert Einstein, David Bohm, Carl Sagan, Alan Watts**

This book's expedition into the essence of the existing attraction network in natural areas explores our central culture's space/time relationships. This is like a movie film because the information on each of this book's pages is like a momentary, separate universe movie frame of a natural area relationship that simultaneously continues the whole movie story on the screen.

Revolutionary Wisdom activates our 54 natural attraction senses to help us address the source of today's nature-prejudiced dilemma[37]. The latter excessively detaches, on average, over 98 percent of our senses, time and thinking from the beneficial, self-correcting ways of Nature, in and around us.

Industrial Society makes us suffer **"Earth Misery"**[5] a runaway, destructive, natural resource, species extinction and mental illness calamity that affects our body, mind, and spirit. This collapse deteriorates personal, social and environmental well-being by ignoring the natural world's self-correcting wisdom. Normally, without telling stories or producing garbage, moment-by-moment the organic life of Nature's intelligence attractively creates Earth's pure optimums of life, diversity, cooperation, balance, beauty, and well-being. Our human being contains every one of them. *(Grok this now, discover what Nature "says" and continue this practice after each paragraph when appropriate).*

Five decades of all-season Organic Psychology research into the restorative essence of natural areas have validated the therapeutic benefits of engaging in a Revolutionary Wisdom Unified Field process. It strengthens any endeavor by scientifically substantiating sensations and feelings as undeniable, self-evident "facts of life" that help us enjoy the happiness and health of whole-life

---

[37] http://www.ecopsych.com/prejudicebigotry.html

accuracy. The art of its hands-on, core learning process empowers us to create sensory natural area moments that let Earth teach and heal.

Where is your life with regard to the information, above? How does it affect your past and present, your dreams and future?

Natural Truth

Obtain real "facts of life" from your real-time, true to life experiences in Nature.

> *Has there ever been a moment when you felt the profound peace in a sunrise? Do you find the sound of waves rippling gently against the lakeshore soothing? Do you feel cheered by the voice of a songbird?*

Think of at least one good experience or moments that you have had in nature: backyard or backcountry; mountain, forest or field; brook, ocean or shoreline; pet, garden or aquarium.

- Try to remember colors, sounds, aromas, music, textures or flavors that might have been part of the experience.
- Did your contact with nature contain comforting emotions or feelings of community, trust or place?
- Did you feel this visit was enchanting, organic or spiritually pleasing?
- Was it supportive, peaceful or healing? Did you feel more balanced or alive?
- Did you sense a personal or natural area euphoria or rapture?
- Did you feel renewed or purified, or that you were part of a greater whole or being?
- Did you feel you belonged?

Can you sense an inner essence of the life of Nature by remembering these experiences without words being part of your thinking, by just how you sensed it? *How did this totality feel?* When you experience it, you begin to know Nature, in and around you, as it knows itself. We have lost contact with this part of our authentic selves because in recent geologic times, as humanity, we came into being with our unique ability to label things and make them into stories. ***Before then, the life of Planet Earth's eons Grokked itself.*** *It was wordless,* including the omission of the words "expert," "profit," "lawyer," "God," "distort" "conquer" "money" and "politician."

> **Every part of wordless Earth held itself together or was pulled together,** moment-by-moment (**m/m**) by attraction. We know this because that is what attraction still does, as it did in the original, beginning of time, as a Unified Field that has been identified by

physics[38]. That same attraction is found in the life of ten billion old stars, in natural areas and our body, mind, and spirit right now.

Today, we call Nature's m/m, story-less, organic, people-absent, self-correcting attraction relationships a "natural area," "wilderness," or the "Web of Life." **We learn to deny that the essence of its life is us devoid of our ability to relate via stories.** *(Grok this now. What does nature "say"?)*

As human's m/m, we can sense, feel, observe and Grok natural areas a balance of life, diversity, beauty, purity, peace, wellness and cooperation that does not result in today's runaway garbage, stress, and abusiveness. We should be able to produce the same in our civilization because we have inherited that know-how. It is in our genetic makeup since Nature loved us into being its quality of life that is also our life. *(Grok this now. What does nature "say"?)*

Scientifically, the closest wilderness to us is, wordless, within us, the most profound love that we can feel.

Long before our appearance Nature organized, corrected and reproduced itself into being what it is *now.* Each TN moment includes the beginning of its life as our planet for all things are attached to each other and communicate through this eternal love of life.

The m/m life of our planet Mother Earth was a pristine attractive natural wonder until humanity, *and our nature-disconnected stories* arrived and deteriorated it as well as warped the mind of our natural-born selves.

From sub-atomics to weather systems and beyond, in scientifically valid, organic sanity, Earth's unadulterated web-of-life still operates, moment by moment, so that everything there is attractive and belongs with no waste or garbage produced. A deep attraction to the whole of life is shared by and for the life of each thing, including us. We call this attraction our love of life or survival. *(Grok this now. What does nature "say"?)*

We are each born a wordless, attractive, natural human infant whom m/m is influenced or socialized by human stories and their long-term effects, constructive or destructive[39].

---

[38] http://www.ecopsych.com/journalcopernicus.html

[39] http://www.ecopsych.com/saneearth.html

The therapeutic process of Organic Psychology holds the natural truths, identified above, to be self-evident. When we visit a natural area quietly, we experience them in the immediate, ever-changing, space/time life of our planet as connections with Nature that are well worth Grokking and repeating. People say that they did not need an instructor, class or book to teach them how to make these contacts. They were and are valid, spontaneous, innate, felt-sense facts. Many folks fondly remember them as wonderful childhood experiences. These reasonable, multiple sensory connections are the unifying heart of revolutionary wisdom. *(Grok this now. What does nature "say"?)*

Where is your life with regard to the information, above? How does it affect your past and present, your dreams and future?

A Self-Evident Fact: *There is no repulsion or negativity in Nature, it is attracted to be.*

#1 An Expedition Challenge

If you have not had a nature experience, as described above, and would like to explore its possibilities for you, go to the most attractive natural area available and Grok it. Sit quietly there for 10 or more minutes just sensing the area.

Shut off the m/m stories buzzing in your mind by repeating the word attraction. "Attraction" correctly labels your time and the things in the area while you are there because you were attracted to have this experience. Also the existence-experience of everything there is held together by immediate attraction. Attraction is the energy that enables you to Grok.

Then spend another 10 minutes or more seeking to enjoy attractive new things there as you become conscious of their attractive m/m calling. Sensitivity to them was already inborn within you. Remember: geologically nature's existence as the life of Earth preceded humanity by 5 billion years and Earth's life is also your life. As verified by your genetic makeup, you are Grokking your origins, and they are present now in this time and space of our Universe.

> **Are you aware of this scientific secret?** The attractive natural area that you visit is as far as its space/time life has come to exist this moment[38] and its life is you, you hold it in common at this moment, this is when the essence of all things is one. As already mentioned, the major difference between the area's m/m life and your life is:
>
> - Its life is silent/mute/non-literate/story-less.
> - Your life's sense of literacy (sense #39) naturally can think, speak, know and relate through abstract stories.

Revolutionary wisdom teaches you the art of scientifically choosing or creating stories that Grok and enhance the life of Earth and yourself. They replace today's negative stories that abusively deteriorate the natural world, in and around us, that also is us.

> **"I go to nature to be soothed and healed and to have my senses put in order." - John Burroughs**

Additional Information:
http://www.ecopsych.com/2004ecoheal.html
http://www.ecopsych.com/54testimonials.html
http://www.ecopsych.com/survey.html

Final exam question: *Why is it significant to use math/science to make sense of the whole of life? (who, what, where, how, when)*

Journal Your Wisdom

It is worth repeating here that as humans we can sense, feel and observe in wildness the immediate wise balance of life, diversity, beauty, purity, peace, wellness and cooperation that does not produce today's runaway garbage, stress, and abusiveness in humanity. We can m/m learn to do the same in our personal lives since scientifically the life of Nature is our life.

We must learn how to let nature's wisdom help us correct or disengage from our nature-disconnecting stories since our attachments to these falsehoods make us produce our abusive personal and global conflicts.

> *IMPORTANT NOTE:* The response box, below, appears throughout this book ***because it enables you to influence your destiny favorably***. The more time and energy you dedicate to completing its instructions **the more you will improve all your relationships and help others do the same.** To be a whole person your story way of knowing must be as close as possible to how your sensory wisdom recognizes itself in a natural area. This avoids conflict, stress, and duality. Although the box directions are usually identical, **each moment in a natural area that you apply them is unique.** It is by connecting with this reality that you best benefit from the self-correcting, 54-sense essence of life that you and the natural world hold in common.
>
> **FOR YOUR CONVENIENCE** the Attraction and Senses list[40] and the **Journal Response Form** in the response box, below, are available in Appendix A: Our Fifty-Four Natural

---

[40] http://www.ecopsych.com/Fifty-four Sense Wisdom.docx

Senses and Sensitivities and Appendix B: Revolutionary Wisdom Response Form in this book.

EXPLORE NOW: VALIDATE. *CONNECT THE CONCEPT, ABOVE, WITH A CONSENTING NATURAL AREA.*
• List the attractions and senses[40] that you find or that find you. (Appendix A: Our Fifty-Four Natural Senses and Sensitivities) • Add this process to an experience you have had in your special area of interest: Art, Creative Writing, Music, Yoga, Parenting, Recovery, Addiction, Renewal, etc.
• **Supportive Reading and Activities:** in a natural area read the first chapter or appendix in *Reconnecting with Nature* including doing its activities and journaling them.
• **Journal Response Form:** For personal growth and reference to describe the value of this section, please complete the Appendix B: Revolutionary Wisdom Response Form: Journal Response Form[41].

REMINDER: The authors learned much of the information in this book from direct contact with attractions in natural areas and nature-loving people. For this reason, after most of the paragraphs here you will be able to find, Grok and validate the truth of the message of the paragraph in the natural area you are connected to. **This Grokking experience *is required of all readers* who seek the credit and special advantages that this course provides.** It should be journaled so that it is available in additional course activities and your application of Organic Psychology to your unique interests and livelihood.

Organic Anchor F: Purpose, Instructions, and Consent

2017 A.D.: The Health of Planet Earth and Humanity

Evidence-based facts and feelings demonstrate that since the beginning of time the pure and balanced life of Nature and Planet Earth has had the wisdom to create and sustain its m/m time/space perfection and in recent geological times, this includes humanity as part of the process. All other scientific "facts of life" today stem from this truth.

We produce our troubles because the story of our scientific world educates us to excessively disconnect from and conquer the health and sanity[9] of the life of Nature rather than embrace it

---

[41] http://www.ecopsych.com/54form.docx

in balanced ways.

The art of mathematically untainted observation of our planet over the centuries shows that **humanity lives inside the life of Planet Earth and its biosphere. Earth's life flows through us continually.**

Every few years all the atoms and energies that we consist of are being replaced by new atoms and energies from the planet. We are it; it is us. Our bodies consist of ten times more cells of other species than our species. Almost thirty percent of our genetic makeup consists of plant genes. All of our atoms are recycled, and some may have been part of the life of loved ones or profound individuals thousands of years ago.

> **"There are as many atoms in a single molecule of your DNA as there are stars in the typical galaxy. We are, each of us, a little universe."-Neil deGrasse Tyson**

We and our planet are identical. We are personifications of the life of the Universe, Nature, and Earth with the exception that we have evolved a unique sense of storytelling literacy. Its contribution to diversity and supporting the life of Earth is that enables us to symbolize and speak our experiences, thoughts, and feelings. This gives us the survival advantage of being able to speak and hear stories so our consciousness can make conceptual maps that we can evaluate and can guide us. We enjoy the survival advantage of storytelling that enables us to survive as part of Earth's life and web-of-life.

A destructive reversal of our life-supportive relationship with Earth today plagues us and produces our disorders. It exists because we have excessively disconnected our stories from connecting with and supporting the self-correcting life of Nature. This has given us the authoritative technological power to manage Earth's life and survive outside the tropics where we biologically and culturally evolved. We are now destructively attached or addicted to using nature-altering power, and so we can't stop. *(Grok this now. What does nature "say"?)*

Where is your life with regard to the information, above? How does it affect your past and present, your dreams and future?

A potential final exam question: *Why is it important to sense that we live in, not on, Planet Earth? (who, what, where, how, when)*

> **A Self-Evident Fact:** We are personifications of the life of the Universe, Nature, and Earth.
> **A Supportive Observation:** *Show me a person who thinks they are not a personification of our vast Universe and I'll show you someone whose thinking is half-vast.*

Mission

In conjunction with a natural area, backyard or backcountry the mission of Applied Organic Ecopsychology and Ecotherapy is to help us:

- Grok the life of Nature and Planet Earth to produce unadulterated, evidence-based, sensory relationships with it that therapeutically increase the well-being of our life, economics, and environment.

- Support Grokking nature in the following areas to reduce their expenses and corruption:

• Education, therapy and mental health
• Physical health, medicine, life and health insurance
• Parenting, welfare, and social services
• Personal, social and international relationships
• Environmental protection, sustainability, and climate change
• Individual, family and spiritual relationships
• Politics, industry, and employment
• Community development and cooperative relationships
• Nature-deteriorating philosophies and institutions
• Imprisonment and military service
• Weddings, meetings, and funerals

***Revolutionary Wisdom* empowers you to achieve its mission** by including in its nature-connecting process the natural area information and activities in its companion books by Michael J. Cohen online or on Amazon.com:

⇒ *Reconnecting with Nature,*

⇒ *Educating, Counseling and Healing with Nature: The Natural Systems Thinking Process webstring model (NSTP)*[11],

⇒ and *With Justice for All*[6].

These books are used to help you keep connected with a natural area while reading this book. This makes the information here like an expert human guide who accompanies you on an expedition, interpretative walk or meditation in a natural area.

Additional information and support are available in the author's other books:

- ⇒ *Well Mind, Well Earth,*
- ⇒ *Connecting with Nature,*
- ⇒ *Nature as Higher Power,*
- ⇒ *Einstein's World: The Natural Systems Thinking Process (NSTP),*
- ⇒ and *Prejudice Against Nature.*

Courses and Degree Programs are offered online that accredit your participation in this book's expeditionary learning process. For this reason, as above, "A Potential final exam question" appears after some sections, and all of them are in Part Four. Visit http://www.projectnatureconnect.org for program details.

EXPLORE NOW: VALIDATE. *CONNECT THE CONCEPT, ABOVE, WITH A CONSENTING NATURAL AREA.*
• List the attractions and senses[40] that you find or that find you. (Appendix A: Our Fifty-Four Natural Senses and Sensitivities) • Add this process to an experience you have had in your special area of interest: Art, Creative Writing, Music, Yoga, Parenting, Recovery, Addiction, Renewal, etc.
• **Supportive Reading and Activities:** in a natural area read the first chapter in *Reconnecting with Nature* (or approved course book) including doing its activities and journaling them.
• **Journal Response Form:** For personal growth and reference to describe the value of this section, please complete the Appendix B: Revolutionary Wisdom Response Form: Journal Response Form[41].

Revolutionary Wisdom enables you, moment by moment, to fully live in the Standard, space/time Universe's real world of timelessness where all things, including love, only exist in each immediate moment[42]. It empowers you to create moments that let the self-correcting life of Earth bring to mind wisdom you inherently know already because you were born with it as part of Nature and it is present now. By applying this knowledge, like an avatar or ambassador of Earth, you can help others benefit by creating similar moments for themselves and others.

---

[42] http://www.ecopsych.com/truthlist.html

To achieve this expertise, in this book's Journal Response Page[41] you write your thoughts feelings and reactions to each section and its natural area Grokking activities as requested. That information m/m comes directly from nature. It is the wisdom you and I are often missing that makes us produce and suffer our problems. You will become stronger just by writing it and discovering more in the process. Once written, you can use it as a learning reference to help yourself and others. It can also be included in social network postings, coursework responses, essays, articles, books and your thesis or dissertation for a degree. In concert with all your 54-senses, it is unique and powerful self-evidence that m/m is scientifically indisputable for your life since you experienced it. It is the missing whole-life information[43] that we each need to solve the problems that we produce by omitting it. Place it in the *revolutionary wisdom closet* you built in your mind.

Reminder: Your Revolutionary Wisdom Safe Deposit Box

In your mind, you have built a strong, empty, sterile storage box or closet whose door only you can open. Each time you discover or Grok a fact of life of Nature/Earth from this book that you know is true for you, lock it in that box, so it is not adulterated by falsehoods or things that are not right for you.

---

[43] http://www.ecopsych.com/interview.html

PART TWO

Implementing What Our Common Culture Already Knows

"I soloed on this activity in a favorite area overlooking the beautiful vista of pinons, ponderosa, juniper, yucca plants, hills, and distant mountain peaks. The day was warm and breezy, birds chirping. I closed my eyes and began to explore the area. I felt plants, rocks, trees and tasted a few as well. The scent of vanilla from Ponderosa pines filled my senses. I felt very relaxed and soon lay down on my back, solidly contained and supported by the planet mother. I felt love and beauty, alert and finely attuned to my surroundings. Birds, insects and the wind all sang their distinctive songs. I felt unique spongy textures, textures I would not have even imagined using sight alone. I heard the wind and traveled with it. So many senses intensely came to life: trust, community, place, gravity, compassion.

Before doing this exercise, I was plagued with all sorts of projects I had lined up for the day. I felt overworked, overwhelmed and somewhat out of control. But I did force myself to move away from these stress-related chores and go into nature. Soon my stresses began to shed, and I started to relax and felt welcomed- as if coming home to the Self. A Self that feels whole when in contact with Nature."

"The scientific method that makes contemporary life possible recognizes that everything only exists in the moment, including this statement, my garden and me."

- Stacey S. Mallory

This book's expedition into the essence of attraction in natural areas, backyard or backcountry, explores our central culture's space/time relationships that are like a movie film. The information on each of its pages is like a momentary, separate universe, real-life movie frame that simultaneously continues the whole movie story on the screen.

Revolutionary Wisdom activates our 54 natural attraction senses to help us address the source of today's nature-prejudiced catastrophe[5]. The latter excessively detaches, on average, over 98 percent of our senses, time and thinking from the beneficial, self-correcting Organics of Nature.

Industrial Society makes us suffer "Earth Misery"[5] a runaway, destructive, natural resource, species extinction and mental illness calamity. It deteriorates personal, social and environmental well-being by ignoring the natural world's space/time wisdom.

In summary, without telling stories or producing garbage, moment-by-moment the life of Nature lovingly creates Earth's pure optimums of life, diversity, cooperation, balance, beauty, well-being and us. We inherit and personify these attributes.

Five decades of all-season Organic Psychology research into the restorative essence of natural areas have validated the therapeutic benefits of engaging in our "GreenWave Unified Field" process. It strengthens any endeavor by scientifically substantiating sensations and feelings as self-evident facts of life that help us enjoy the happiness of whole-life accuracy. The art of its hands-on, core learning process enables us to create sensory natural area moments that let Earth teach and heal.

From our scientific perspective, here is our hypothesis for writing this book/course: "Reading Applied Ecopsychology information while 54-sense connected with authentic natural areas empowers us to learn and teach others to build true, evidence-based, heart-centered relationships that are fused to the self-correcting ways of the life of Earth and its beauty."

"There is no other door to knowledge than the door Nature opens. And there is no truth but the truth we discover in Nature." **- Luther Burbank**

GAINING CONSENT: CREATE NATURE-BLENDED MOMENTS

Wild Deer and Moose have been observed walking by and through flocks of Canadian Geese, towering above them, yet when a human approaches the geese move away. This is true of most wildlife. They know from experience and, maybe genetically, that humanity is a danger to be avoided. For this reason, Organic Psychology does not work correctly, and it can backfire if your personal life does not first establish a trusting relationship with the thing in the natural area you want to connect with and learn from cooperatively.

#2 An Expedition Challenge

Section One. An Expedition Metaphor:

> Bob: What would happen if you walked past a complete stranger and into his house, opened his refrigerator door, took out his sandwich and ate it?
>
> Eileen: He would be frightened, upset and angry. A fight might break out, or the police called.
>
> Bob: How could that disruption be avoided?
>
> Eileen: You could first make a friendly connection with the stranger, thank them for listening, communicate your desire for food, and ask for and obtain his consent for you to enter the house and satisfy your hunger. A respectful friendship might develop.

1. Nature enables things to build balanced relationships through natural attraction energies. **Notice how you feel right now, then go to something in nature, first A) that attracts you, or second B) that you choose because you find it attractive or interesting. A park, backyard, aquarium, even a pet or potted plant will do.** Their attractiveness is a tangible sensory connection, some of our 54 natural senses being activated. It invites, welcomes and consciously, feelingly connects you to them.

Just like thirst naturally attracts you to water, or contact with water may make you thirsty, you are biologically built to connect with the Earth community through cohesive sensations naturally, natural "webstring" web-of-life attraction loves that can't tell stories but feels good. The more natural and attractive a natural area or the thing is, the more worthwhile the results of this activity. A goldfish or a flower may be better than a wilderness area if it is more attractive to you.

2. **Thank the natural attraction that brings you to this area for being there for you.** Thank it for safely activating a good feeling in you through this attraction connection.

3. Recognize that as part of the life of Earth community, justifiably, as in your life, this natural area or thing desires and has a right to safely exist, build beneficial relationships and grow, just as

you do. **Decide that you are going to respect its integrity by asking for its permission to visit it.**

4. Because our story way of life socializes to think and act in nature-disconnected or conquering habits, we are foreigners to the life of this area. Its life fears us in its way. We build relationships through stories, accurate or not, that it does not understand and that often hurt it while ignoring that it is also us, so we are producing our disorders.

Silently, aloud or in writing, respectfully ask this natural area for its consent for you be there and do this activity there. It will not permit you to visit if you are going to injure, destroy or defame it, or if it will not be safe for you. Remember, in nature, negative relationships are not attractive. **Promise this area that you will treat it honorably** because you love your life (to survive) and it is Earth's life and vice versa.

5. **Sense the area for 7 seconds or more in silence and respect. Be aware of negative signals from stress, discouragement or danger from it, such as thorns, bees, poison plants, ticks, cliff faces or unpleasant memories, thoughts or feelings.** If they appear, thank them for their "attractive" message to help you find more attractive ways to obtain good feeling and rewards safely.

[FOR EXAMPLE: "Our group was asked to select something attractive, sight unseen, from a bag full of miscellaneous objects. One adult woman blindly selected a piece of wood in the bag because she was attracted to its shape and smoothness when she groped and explored it by touch. However, she had a negative reaction to the wood once she took it out of the bag and saw it. At first, she did not know why she did not like it when she viewed it, but in time, perhaps through her dreams, she realized it was a subconscious reaction. The wood was the same shade of blue as the walls of a room where, as a child, she had been molested. Ordinarily, during the 7 second waiting period in a natural area, another attraction would have appeared to her if she could have seen the color of the stick."]

> A. **When the 7 seconds are up, note if the area still feels attractive, or has become more attractive.** If either, it has consented to your visit through a multitude of your natural senses. Proceed to 6.
>
> B. If this part of the natural area no longer feels attractive or is replaced by another attraction, **thank it for its guidance and simply select another natural part of the area that feels attractive to you.** Then repeat the gaining permission process. Do this until you find a 7 second period when a safe attraction feeling remains for a place, color, shape or another natural thing. When this occurs, you have 54-sense permission to visit it. In

that safe moment, many additional natural senses are happily connecting, consenting and blending to the point that they energize onto your screen so consciousness so you can experience them.

6. As soon as you gain a natural attraction's permission to visit, **genuinely thank it for giving its consent.** You might give it a gift it would like such as compost or exhale on a plant leaf there.

7. Now: **Compare how you feel about being in this mutually supportive moment with how you felt when you first started doing this activity.** Has any change occurred because you gained the life of this natural area's consent and thanked it for consenting? Does the area feel better or friendlier to you? Do you find it more attractive, ethical or rewarding now than before you received its consent and thanked it? Do you feel better about yourself, more supported by the life community? Do you feel less stressed?

Write down what occurred and if you obtained good feelings or rewards from doing this activity, what they were and whether you trust them. See if you can identify which of you 54 natural senses were involved (See Appendix A: Our Fifty-Four Natural Senses and Sensitivities). Share this information with people close to you or others who are doing the activity.

If you find that thankfully gaining permission to visit the life of a natural area is rewarding, remember that whenever you want to feel rewarded and less stressed, you can repeat this activity.

Remember too, that the life of nature exists in people, it is our life, and your sense can connect with its attractive 54-sense existence at any moment.

Thankfully asking permission and gaining consent to relate to the life of people's inner nature also provides rewards and helps build good social relationships. It is also satisfying if you thankfully request that people seek permission from you concerning how they relate to the life of nature in you. Doing this activity may help them learn to relate that way.

Learn to trust the process and sensations in this consensus experience because they are safe, supportive, earth-linked, sustainable, in balance, intelligent and they feel good. In them lies hope.

Section Two

To complete this activity, apply what you have learned here by always seeking consent to visit members of the web-of-life's plant-animal-mineral-energy community.

Below are some reactions activity participants have shared with each other.

"It was hot. Soon after I asked for permission to visit with the grove of young trees, a gentle, refreshing breeze came through them. It cooled me, and the trees waved their leaves at me. It felt good like the grove smiled its consent. Thanking the grove strengthened that feeling as does share the experience with you now."- Anonymous

"I was attracted to the sound of a raven on the rocks ahead. I stopped and sought its consent for me to enjoy its presence. It began to come closer and closer, increasing my delight and excitement. That was so fun and unforgettable. I feel thankful for that experience and this group." - Anonymous

"My whole attraction to the moss on the rock increased. I felt more intensely than when I first arrived, it felt like a hug from the planet." - Anonymous

"Many times I have forced myself to back away from the deadlines and details of our super demanding lives and return to nature. In every case, I have found the same welcoming feeling of self. In fact, I had to stop today (a particularly stressful day) and gain permission to connect from a beautiful maple tree outside of my office window whose leaves are just popping out of their buds. People often ask me how I stay so calm while they are all running around like crazy. When I try to share the ideas of this activity, so many people look at me as if I was the crazy one." - Anonymous

"I was attracted to the view of my little town from the water tower and especially the green pockets of nature in it that I'd never noticed before. This led me to spend more time with them and with the people I met there. I formed naturally attractive relationships with some of them that ordinarily, I would have ignored." - Anonymous

Now that you have learned how to create moments that make contact with the life of Nature and let Earth speak you can use this process to help you recognize the accuracy of these two accepted facts, repeated from above

On average, in contemporary society, over 95 percent of our time and 99 percent of our thinking and feeling are excessively disconnected from the self-correcting way that the life of Nature works in and around us. Instead, we spend our time in homes, schools, vehicles, and business or entertainment buildings while connecting through stories and media that conquer the life of Nature/Earth. It does not understand our stories because it neither speaks, reads nor writes its ways yet naturally create balanced optimums of life, diversity, beauty, and cooperation without producing our runaway garbage, stress, and abusiveness.

A potential final exam question: *What benefit did you find from doing the "Gaining Consent" activity? (who, where, how, when, why)*

EXPLORE NOW: VALIDATE. *CONNECT THE CONCEPT, ABOVE, WITH A CONSENTING NATURAL AREA.*
• List the attractions and senses[40] that you find or that find you. (Appendix A: Our Fifty-Four Natural Senses and Sensitivities) • Add this process to an experience you have had in your special area of interest: Art, Creative Writing, Music, Yoga, Parenting, Recovery, Addiction, Renewal, etc.
• **Supportive Reading and Activities:** in a natural area read the next chapter in *Reconnecting with Nature* (or approved course book) including doing its activities and journaling them.
• **Journal Response Form:** For personal growth and reference to describe the value of this section, please complete the Appendix B: Revolutionary Wisdom Response Form: Journal Response Form[41].

MAKE A WARRANTIED FACT CHECK: Visit the warrantied fact list http://www.ecopsych.com/54warrantfact.docx and insert dates on facts that you are satisfied you have learned or know to this point in the book.

If you are aware of the natural world that you are visiting while reading this book, your 54-senses consciously or sub-consciously blend nature's sensory wisdom with what you are reading. This is because you have connected your consciousness with Nature's ways in this natural area. Both the words and Nature register on your sense of consciousness (sense #43). This helps your sense of reason (sense #42) search for additional sensibility and become 45 ways more sensible and sensitive about the value of the web of life's organic whole of life and its peace and wellness that includes your personal life.

You may already have experienced the above when you had spent quiet time in a natural area or sought it when it was needed. This means that it is a self-evident fact of life for you. Do you recognize that quiet time in nature is far more than just getting away from your problems?

When you read something in this book, it asks you to reinforce it by discovering how your senses register it in the natural area where you are located. Start exploring this connection now. Can you sense or feel examples of what you have just read in a natural area? Use Appendix A: Our Fifty-Four Natural Senses and Sensitivities to help you.

Additional Information: http://www.ecopsych.com/54transformation.html

A potential final exam question: *Do you think your contact with nature while reading this book is making a difference? (who, what, where, how, when, why)*

EXPLORE NOW: VALIDATE. *CONNECT THE CONCEPT, ABOVE, WITH A CONSENTING NATURAL AREA.*
• List the attractions and senses[40] that you find or that find you. (Appendix A: Our Fifty-Four Natural Senses and Sensitivities) • Add this process to an experience you have had in your special area of interest: Art, Creative Writing, Music, Yoga, Parenting, Recovery, Addiction, Renewal, etc.
• **Supportive Reading and Activities:** in a natural area read the next chapter in *Reconnecting with Nature* (or approved course book) including doing its activities and journaling them.
• **Journal Response Form:** For personal growth and reference to describe the value of this section, please complete the Appendix B: Revolutionary Wisdom Response Form: Journal Response Form[41].

NOTE: If you want credit for engaging in this book's natural area expedition now or later, be sure to write down attractive things in nature that you sense and feel as you read it or look forward to re-doing this later, as it will be required.

Process Importance

Nature-connecting activities, like those above, are significant. This book enables you to use them to scientifically experience and include the life of Nature/Earth in determining the accuracy and value of a story or relationship. They enable you to work with scientifically, whole-life facts that include the wisdom of the way the life of nature works. The activities are always available to you so you can use them to help you deal with problems and disorders in your life. In some programs, this is recognized as using Nature as Higher Power.

Make yourself a self-supportive playground. Involve yourself in this book slowly while in a natural area so you can learn the area's "language." Give yourself the time and space to do linked activities and read the linked pages. Expand and fortify what you find attractive in this nature-reconnecting process.

Enjoy at least one night's sleep before doing a new activity. That allows nature to produce new paths in your mentality from your activity experience while your story world is asleep so it cannot block or impair your experienced, nature-connected truths.

You can read the entire book in a few hours, but then its information will make little impact on your habitual ways of thinking and relating. For this reason, imperative gains are unlikely.

To use revolutionary wisdom to lead a more sensible and rewarding life, **it may take 4-24 hours to complete one chapter including its activities and links.** They can be spread out over whatever time you need.

Additional Information: http://www.ecopsych.com/54transformation.html

A Potential final exam question: *Why is it important to get a night's sleep between activities?*

EXPLORE NOW: VALIDATE. *CONNECT THE CONCEPT, ABOVE, WITH A CONSENTING NATURAL AREA.*
• List the attractions and senses[40] that you find or that find you. (Appendix A: Our Fifty-Four Natural Senses and Sensitivities) • Add this process to an experience you have had in your special area of interest: Art, Creative Writing, Music, Yoga, Parenting, Recovery, Addiction, Renewal, etc.
• **Supportive Reading and Activities:** in a natural area read the next chapter in *Reconnecting with Nature* (or approved course book) including doing its activities and journaling them.
• **Journal Response Form:** For personal growth and reference to describe the value of this section, please complete the Appendix B: Revolutionary Wisdom Response Form: Journal Response Form[41].

PART THREE

Stories About Why Organic Psychology Works

“We don't exclusively own our natural attraction web-of-life senses and sensitivities. Rather, they are a voice we share with every species and mineral. They attract the natural community, the web of life, to beneficially flow through us and us through it, atom by atom, moment by moment. That makes earth like our other body, our second mother. This is absolutely one of my favorite aspects of this course. It comes closest to expressing for me the sheer reverence in being alive. How is this possible? We are so separate and yet simultaneously irrevocably of and within our environments! Sheer genius. I appreciate being able to see this and express it in terms that strengthen it.

All my life I have been able to slip easily into feeling one with life, with the nameless, natural attraction, essence of nature, what many perceive as God. And to be able to share this knowledge through these words is such a gift. It conveys what those moments of unity and peace feel like, that my molecules are in exchange with all of life and at one with all of life and that is natural and eternal. This moment is great natural and eternal; I can think in this way forever if I want to make nature appreciate how I care about it."

"To find the right road, in reality as well as in imagination we must return to our origins and from them go forward again in a truly civilized, not a merely artificial, way of life."

- **Michael J. Cohen,** Expedition Education[44], 1968

Eight overview stories that may help you understand and convey to others the how what and why of the Organic Psychology process.

Story 1. The Earth Egg Scenario

In imagination and metaphor, in this story, we safely live *inside the life* of an otherwise **empty** Earth egg. The eggshell is our planet's surface as seen from the moon. In this story, its life, consisting of its thin shell, is held together at any given moment by natural attraction relationships that want to survive and attractively continue their life as the eggshell.

Moment by moment the shell is attracted to grow itself by fulfilling new diverse attractions that it finds have to value. Its space/time life has evolved and sustained itself and the life of the egg this way since its beginning.

The thin life of the egg and eggshell is non-literate. Unlike humanity, it cannot tell or understands stories and labels.

Inside the egg, if our stories attract us to connect with the shell in mutually supportive, sensory ways, moment by moment the life of the shell nurtures and protects us as part of the life of the egg growing itself and us into the next moment.

When our stories injure or disconnect us from the life of the shell, devoid of our support, part of the shell breaks. We are vulnerable and hurt by this the void this break makes and so is the egg. It feels like our life has been abandoned, we feel anxious or grief.

Stories that are about our life **within the space** of the egg are our past experiences and knowledge. They do not connect us at the moment with the life of the shell and information it needs because the eggshell does not understand stories. Instead, they often make or leave disconnection holes that injure our life and the egg's life.

---

[44] https://www.amazon.com/Our-classroom-wild-America-encounters/dp/B000721C1W

Stories that are about our lives **outside the space** of the egg are our predicted future experiences and knowledge. They do not connect us at the moment with the life of the shell because the shell does not understand stories. Instead, they often make or leave unproven or outdated disconnection holes that injure our life and the egg's life.

Sensory attraction experiences and stories that at the moment **make genuine, supportive contact with the shell through our 54 natural senses**[27] increase its and our well-being and happiness. This attraction connection also helps us seal the injured holes our disconnected stories have previously made.

> **The egg is not empty.** It is filled with and consists of all the other members of Web-of-Life plant, animal, mineral and energy kingdoms held together individually and collectively by the Unified Field natural attraction of the moment. Our lives consist of being part of space/time life of the egg growing more attractive along with us.
>
> Social relationships that do not close the holes made in the shell help the holes reinforce themselves and grow as holes as time/space proceeds.

Where is your life in this story? How does it affect your past and present, your dreams and future?

A potential final exam question: *What makes holes in the eggshell and what fills them? (who, where, how, when, why)*

EXPLORE NOW: VALIDATE. *CONNECT THE CONCEPT, ABOVE, WITH A CONSENTING NATURAL AREA.*
• List the attractions and senses[40] that you find or that find you. (Appendix A: Our Fifty-Four Natural Senses and Sensitivities) • Add this process to an experience you have had in your special area of interest: Art, Creative Writing, Music, Yoga, Parenting, Recovery, Addiction, Renewal, etc.
• **Supportive Reading and Activities:** in a natural area read the next chapter in *Reconnecting with Nature* (or approved course book) including doing its activities and journaling them.
• **Journal Response Form:** For personal growth and reference to describe the value of this section, please complete the Appendix B: Revolutionary Wisdom Response Form: Journal Response Form[41].

Story 2. The Remedy for Blind-ness

Contemporary people, being blinded to 48 of our natural senses, today, are no different than the blind Governing Council of an island society that once consisted entirely of non-sighted people. Citizens on that island were content and adequate in their uniquely adapted ways, even though an incurable disease had numbed each person into blindness by the age of two years.

One day, Gulliver, a shipwrecked castaway, half-dead, washed up on the island. Compassionately, many community members of the blind society nurtured him to full recovery. Then, because he was sighted, he became the bane of their existence. He demanded things unknown to them in their blindness: windows, lights, books, television, painted colors, and sunglasses.

The Governing Council investigated and discovered Gulliver's trouble. Gulliver had eyes that could see.

The Council, in their wisdom, solved their problem; they took Gulliver's sight away. They chemically numbed his optical nerves in the same way their early childhood disease number theirs.

Although Gulliver adjusted to his loss, he emotionally hurt whenever he thought about how this island society made him lose his precious sense of sight.

Over time, Gulliver noticed that when he visited natural areas, in compensation for his blindness, many of his remaining 52 natural senses became stronger and more intensely registered in his awareness. This was like the mental renewal provided by a simple walk in the park or along a beach. The contact with nature invigorated him and improved how he thought and felt.

Soon Gulliver invented a science that consisted of activities in nature that helped him strengthen his nature-reconnecting, sensory enhancing process. As he practiced it, his connections to the restorative healing powers of natural systems within and around him increased. Slowly his eyesight un-numbed, recovered and returned.

Gulliver introduced his blind friends to his activities that improved the resilience, energy, and sensibility he had gained along with his delight in enhancing his sight, life, and health. They learned to use his applied science, thrilled to similar results and helped others do the same.

In time, most of the society of the island community regained sightedness and increased their well-being, pleasure, and ability to think in whole ways. In the process, with gratitude and deep passion, they became protective of nature because their innate love for it had been reinforced in their consciousness and rewarded. That love from it had made them well. It made sense, felt right and further helped them heal. They became more happy and healthy.

Where is your life in this story? How does it affect your past and present, your dreams and future? *(Grok this now. What does nature "say"?)*

A potential final exam question: *Is it important to think with advanced science and technology in a natural area? (who, what, where, how, when, why)* **EXPLORE NOW: VALIDATE.** *CONNECT THE CONCEPT, ABOVE, WITH A CONSENTING NATURAL AREA.*
• List the attractions and senses[40] that you find or that find you. (Appendix A: Our Fifty-Four Natural Senses and Sensitivities) • Add this process to an experience you have had in your special area of interest: Art, Creative Writing, Music, Yoga, Parenting, Recovery, Addiction, Renewal, etc.
• **Supportive Reading and Activities:** in a natural area read the next chapter in *Reconnecting with Nature* (or approved course book) including doing its activities and journaling them.
• **Journal Response Form:** For personal growth and reference to describe the value of this section, please complete the Appendix B: Revolutionary Wisdom Response Form: Journal Response Form[41].

Story 3. The Ego's Story: Does your Ego Understand it?

The life of Nature cannot tell stories and although our Ego is our story about who and what we are, its essence and foundation obviously must be, as a story, disconnected from story-less Nature. Our contemporary Ego tells itself a story that omits this disconnection fact while it tells itself how intelligent it is. Its story, however, excludes its destructive impacts on nature, in and around us, as it celebrates it is excessive scientific and technological advances and profits. It is as if its gestalt is money rather than the life of Earth.

The Ego disregards its destructive impacts because it recognizes that it and they are apparently neither responsible nor intelligent. Its life deteriorates nature, the Ego's own life support system. It omits its deleterious effects because **it is embarrassing that it is not even smart enough to know how to stop what it knows how to start. It does not know how to avoid its destructive effects. It does not know how to make food from minerals.**

Out of shame, the Ego's life dismisses that it and its story are part of Nature and that it has put its own life at risk by injuring the natural world. Its story has foolishly disconnected itself from the profound, self-correcting ways of nature's silent life and then attached or addicted itself to its own nature-disconnecting stories and artifacts. It is so short-circuited that depends on their inaccuracy for help to deal with the very problems they are causing.

The Ego, being "literate smart," demeans non-literate muteness, including how nature works, in and around us. This is like letting a mouse manage the website for Amazon.com.
Today's personal and collective Ego cannot solve its short circuit because it is in denial. It denies it has produced a short-circuit error that omits the wise, life-in-balance, self-correcting powers of Nature that could help it remedy its blunder.

As victims of our Ego's flawed, established ways, we are indoctrinated to turn to them, rather than to Nature for help with this problem. *Revolutionary Wisdom* gives us the remedy we need for this dilemma. It provides us with the tools necessary for us to plug our senses directly into the supportive energies of the pure, balanced and self-correcting life of Nature in a natural area, backyard or backcountry. It recognizes that this is reasonable for us to do, no matter what our confused Ego says to the contrary. Our Ego, to be cool, often conforms to the similarly misguided Egos of our neighbors and folks in business suits.

Where is your life in this story? How does it affect your past and present, your dreams and future?

A potential final exam question: *How and why does the Ego mislead itself? (who, what, where, when)*

EXPLORE NOW: VALIDATE. *CONNECT THE CONCEPT, ABOVE, WITH A CONSENTING NATURAL AREA.*
• List the attractions and senses[40] that you find or that find you. (Appendix A: Our Fifty-Four Natural Senses and Sensitivities) • Add this process to an experience you have had in your special area of interest: Art, Creative Writing, Music, Yoga, Parenting, Recovery, Addiction, Renewal, etc.
• **Supportive Reading and Activities:** in a natural area read the next chapter in *Reconnecting with Nature* (or approved course book) including doing its activities and journaling them.
• **Journal Response Form:** For personal growth and reference to describe the value of this section, please complete the Appendix B: Revolutionary Wisdom Response Form: Journal Response Form[41].

Story 4. No Laughing Matter?

A scientist put a Frog on the ground and said: "Jump frog jump." The Frog jumped four feet. The scientist wrote in his notebook, "Frog with four feet, jumps four feet."

The scientist cut off one of one of the frog's legs and said: "Jump frog jump.". The Frog jumped three feet. The scientist wrote in his notebook, "Frog with three feet, jumps three feet."

So the scientist cut off another leg and said: "Jump frog jump." The scientist wrote in his notebook "Frog with two feet, jumps two feet."

The scientist cut off one more leg and said: "Jump frog jump." The Frog jumped one foot. The scientist wrote in his notebook, "Frog with one foot, jumps one foot."

The scientist cut off the frog's last leg and said: "Jump frog jump." The frog did not jump so the scientist said "Frog jump. Frog jump. FROG JUMP!" Nothing happened. The scientist wrote in his notebook, ***"A frog with no feet, goes deaf."***

This story is no laughing matter when you recognize that we are born with <u>54 natural senses</u>, and our scientific *objectivity* cuts us off from the natural world's *subjectivity,* so our intelligence habitually thinks and relates with the <u>nonsense of only eight of the 54</u>. It is like X-ray machine exposure has crippled your computer operating system by 85%. *(Grok this now. What does nature "say"?)*

Another Laughing Matter?

> A small plane ran out of gas while it was flying the smartest man in the world, a pastor, an outdoor educator and the pilot to a conference. The pilot immediately grabbed one of the three parachutes on board and jumped from the plane.
>
> The smartest man in the world said, "The world needs my super intelligence and knowledge," and he grabbed the second parachute and jumped out.
>
> The pastor said to the outdoor educator, "I have lived a long life, and I am prepared to meet the Lord. You take the last parachute."
>
> "Don't sweat it, pastor" replied the outdoor educator, "The smartest man in the world just jumped out wearing my backpack."

Where is your life in this story? How does it affect your past and present, your dreams and future? *(Grok this now. What does nature "say"?)*

A potential final exam question: *Why do think these stories make our sense of reason laugh? (who, what, where, how, when)*

EXPLORE NOW: VALIDATE. *CONNECT THE CONCEPT, ABOVE, WITH A CONSENTING NATURAL AREA.*
• List the attractions and senses[40] that you find or that find you. (Appendix A: Our Fifty-Four Natural Senses and Sensitivities) • Add this process to an experience you have had in your special area of interest: Art, Creative Writing, Music, Yoga, Parenting, Recovery, Addiction, Renewal, etc.
• **Supportive Reading and Activities:** in a natural area read the next chapter in *Reconnecting with Nature* (or approved course book) including doing its activities and journaling them.
• **Journal Response Form:** For personal growth and reference to describe the value of this section, please complete the Appendix B: Revolutionary Wisdom Response Form: Journal Response Form[41].

Story 5. Hidden Reality

April 22, 2000 (After April 22, 2017, increase the figures, below, by 30 percent)

EARTH DAY PLUS THIRTY, AS SEEN BY THE EARTH

By Donella Meadows, the adjunct professor at Dartmouth College. *Used with permission*

If, in the thirty Earth Day celebrations we have held since 1970, (now forty-eight in 2017), the human population and economy have become any more respectful of the Earth, the Earth hasn't noticed.

Fancy speeches do not impress the planet. Leonardo DiCaprio interviewing Bill Clinton about global warming is not an Earth-shaking event. The Earth has no way of registering good intentions or future inventions or high hopes. It doesn't even pay attention to dollars, which are, from a planet's point of view, just a charming human invention. Planets measure only physical things-energy and materials and their flows into and out of the changing populations of living creatures.

What the life of Earth sees[45] is that on the first Earth Day in 1970 there were 3.7 billion of those hyperactive critters called humans, and now there are over 6 billion. *(2013 update, 7.046 billion)*

Back in 1970, those humans drew from the Earth's crust 46 million barrels of oil every day-now they draw 78 million.

Natural gas extraction has nearly tripled in thirty years, from 34 trillion cubic feet per year to 95 trillion. We mined 2.2 billion metric tons in 1970; this year we'll mine about 3.8 billion. The planet feels this fossil fuel use in many ways, as the fuels are extracted (and spilled) and shipped (and spilled) and refined (generating toxics) and burned into various pollutants, including carbon dioxide, which traps outgoing energy and warms things up. Despite global conferences and heroic promises, what the Earth notices is that human carbon emissions have increased from 3.9 million metric tons in 1970 to an estimated 6.4 million this year, 2000, and no end to the increase is in sight.

You would think that an unimaginably massive thing like a planet would not notice the one degree (Fahrenheit) warming it has experienced since 1970. But on the scale of a whole planet, one degree is a big deal, primarily since it is not spread evenly. The poles have warmed more than the equator, the winters more than the summers, the nights more than the days. That means that temperature DIFFERENCES from one place to another have been changing much more than the average temperature has changed. Temperature differences are what make winds blow, rains rain, ocean currents flow.

All creatures, including humans, are exquisitely attuned to the weather. All creatures, including us, are noticing weather weirdness and trying to adjust, by moving, by fruiting earlier or migrating later, by building up whatever protections are possible against flood and drought. The Earth is reacting to weather changes too, shrinking glaciers, splitting off nation-sized chunks of Antarctic ice sheet, enhancing the cycles we call El Nino and La Nina.

"Earth Day, Shmearth Day," the planet must be thinking as its fever mounts. "Are you folks ever going to take me seriously?"

Since the first Earth Day, our global vehicle population has swelled from 246 to 730 million. Air traffic has gone up by a factor of six. The rate at which we grind up trees to make paper has doubled (to 200 million metric tons per year). We coax from the soil, with the help of strange chemicals, 2.25 times as much wheat, 2.5 times as much corn, 2.2 times as much rice, almost

---

45 http://www.ecopsych.com/livingplanetearthkey.html

twice as much sugar, nearly four times as many soybeans as we did thirty years ago. We pull from the oceans almost twice as much fish.

With the fish, we can see clearly how the planet behaves when we push it too far. It does not feel sorry for us; it just follows its own rules. Fish become harder and harder to find. If they are caught before they're old enough to reproduce, if their nursery habitat is destroyed, if we scoop up not only the cod but the capelin upon which the cod feeds, the fish may never come back. The Earth does not care that we didn't mean it, that we promise not to do it again, that we make nice gestures every Earth Day.

We have among us die-hard optimists who will berate me for not reporting the good news since the first Earth Day. There is plenty of it, but it is mostly measured in human terms, not Earth terms. Average human life expectancy has risen since 1970 from 58 to 66 years. Gross world product has more than doubled, from 16 to 39 trillion dollars. Recycling has increased, but so has trash generation, so the Earth receives more garbage than ever before. Wind and solar power generation have soared, but so have coal-fired, gas-fired and nuclear generation.

In human terms, there has been breathtaking progress. In 1970 there weren't any cell phones or video players. There was no Internet; there were no dot-coms. Nor was anyone infected with AIDS, of course, nor did we have to worry about genetic engineering. Global spending on advertising was only one-third of what it is now (in inflation-corrected dollars). The third-world debt was one-eighth of what it is now.

Whether you call any of that progress, it is all beneath the notice of the Earth. What the Earth sees is that its species are vanishing at a rate it hasn't seen in 65 million years. That 40 percent of its agricultural soils have been degraded. That half its forests have disappeared, and half its wetlands have been filled or drained, and that despite Earth Day, all these trends are accelerating.

All these increases come with devastating human, resource, and monetary costs, along with misery factors that keep growing.

Earth Day is beginning to remind me of Mother's Day, a commercial occasion upon which you buy flowers for the person who, every other day of the year, cleans up after you. Guilt-assuaging. Trivializing. Dangerous. All mothers have their breaking points. Mother Earth does not soften hers with patience or forgiveness or sentimentality.

The wholeness of the life of our earth mother dances in and around us. What we do to it, we do to our personal and collective body, mind and spirit.

Ten years later

The Global Footprint Network June 2011-2015

You really do have to wonder whether a few years from now we'll look back at the first decade of the 21st century — when food prices spiked, energy prices soared, world population surged, tornados plowed through cities, floods and droughts set records, populations were displaced and governments were threatened by the confluence of it all — and ask ourselves: What were we thinking? How did we not panic when the evidence was so obvious that we'd crossed some growth/climate/natural resource/population redlines all at once?

We are currently growing at a rate that is using up the Earth's resources far faster than they can be sustainably replenished, so we are eating into the future. Right now, global growth is exploiting about one and a half Earths. Having only one planet makes this a rather significant problem. When you are surrounded by something so big that requires you to change everything about the way you think and see the world, then denial is the natural response. But the longer we wait, the bigger the response required and the higher the misery produced.

If you cut down more trees than you grow, you run out of trees. If you put additional nitrogen into a water system, you change the type and quantity of life that water can support. If you thicken the Earth's CO_2 blanket, the Earth gets warmer. If you do all these, and much more, nature-destructive things at once, you change the way the whole system of planet Earth behaves, in and around us, while producing the misery of destructive social, economic, and life support relationships.

In 1975, Earth Overshoot Day—the approximate date our resource consumption for a given year exceeds the planet's ability to replenish itself—was December 30.

In 1993, Earth Overshoot Day was October 21.

In 2003, Overshoot Day was on September 22.

August 2 was Earth Overshoot Day 2017, marking the date when humanity exhausts nature's budget for the year. We are now operating in an overdraft on borrowed time. Given current trends in our wanting and consumption, one thing is clear: Earth Overshoot Day arrives a few days earlier each year.

Revolutionary Wisdom identifies the point source of this critical dilemma and offers a practical, whole life art and science remedy for it from a maverick genius[23] Our socialization omits this organic antidote because it is inconvenient to our ever-wanting, never-satisfied, excessive conquest and exploitation of nature for profit. Instead, we are trained and paid as perpetrators of this outrage that can be seen as the evisceration of Planet Earth and ourselves[46] by our nature-disconnected stories.

Where is your life in this story? How does it affect your past and present, your dreams and future? *(Grok this now. What does nature "say"?)*

Story 6. Hidden Reality: Seven Blind Leaders and the Elephant

A story concerning seven blind leaders touching an elephant and arguing about what it is conveys the dilemmas of our society's blindness to Planet Earth and our natural senses. In the story, each blind leader describes the Elephant based on what part of her they are touching as they stand by her. While one calls the elephant a pipe (the tusk), another says the Elephant is a snake (trunk) or like a rope (tail), etc. They stop there in confusion. Each goes back to their followers where the senses and feelings of the latter in time addict to the story of their blind leader. The story becomes the dogma of their leader's Institution with a passion that leads to stronger unity in their community. However, this is accompanied by misunderstandings, stalemates, arguments, fights, and wars with the other leader's groups.

The same story as above with a happier ending is that each blind leader, not knowing about the others, follows his or her attraction to learn more about the Elephant. With the elephant's non-verbal consent that she conveys by attracting them to visit her (by attractively dancing), they crawl all over her, discover many new things and meet each other in the process. Doing this exploratory connection "dance," they happily recognize from the experience that they feel healthier. While they are there, each of them is attracted to what is happening including the Elephant; she cooperatively helps them be there. They thank the Elephant for her consent to let blind people explore and learn about her through their senses. They also bond to each other from the delightful meeting and agree that the significant difference they can sense between the Elephant and themselves is that when they speak to each other, the Elephant does not understand what they say, it is non-literate, but far from dumb. When they return to their followers, their group is attracted to them. They also readily unify with other groups and the Elephant, too, for they all hold in common, the unifying Dance of their mutually shared attraction to her and each other.

Attraction is the unifier in this story because, by definition, unify is what attraction/(love) does in word and deed. **To be part of a system, a living being must be in communication and unified with the system.** Otherwise, things trespass each other and deteriorate into toxins and

---

[46] http://www.ecopsych.com/journalmindrape.html

garbage rather than balance into stronger, unified and untainted systems.

The Elephant, Leaders, and Followers are all expressions of Earth's/Nature's Standard Universe system whose dance communicates and unifies in purity by attraction. However, the Leaders mostly communicate by Literacy/Language/Stories that the Elephant neither understand nor uses.

The goal of this expedition course is for our Literacy to learn how to articulate natural attraction so we may communicate with and join the Earth/Nature Attraction Dance to the benefit of all.

We learn how to organically articulate by doing it, by Grokking in natural areas, one supportive moment after another. Each natural area moment makes a safe space for the Elephant along with its "dumb" web of life, to teach us what we need to know. This helps us be better dancers and help others do the same; it increases well-being for all.

Where is your life in this story? How does it affect your past and present, your dreams and future? *(Grok this now. What does nature "say"?)*

Story #7. The Nine-Legged Dog

As I returned home from my hike, the school bus stopped at the corner and dropped off two young girls from first and third grade. We walked down the country road together, and we talked about how they liked school. I made the mistake of asking them if they learned mathematics there and they said they did. Although I soon wished I hadn't, I asked them if they could answer a simple math question that I had seen on a critical thinking math aptitude/intelligence test. This logical reasoning question is used to help determine our mathematical ability or our competency to qualify for the rewards of a better job, higher salary or greater prestige and self-worth. Here's the question:

> 'If you count a dog's tail as one of its legs, how many legs does a normal dog have?'
>
> With smiles, both girls proudly responded "five" and looked at me as if I had asked a dumb question.
>
> 'Yes,' I replied, "And have you ever seen a real dog? How many legs does it have?"
>
> 'Four'
>
> 'Is its tail one of its legs?'

'No, silly,'

'Well then, is your mathematics answer correct, doesn't a real dog only have four legs?"

As they thought about my question, sadness filled the girl's faces. They thought they got the question wrong or that I said they were wrong. The younger girl was about to cry, and tension built in each of us.

I was mortified about what I had done and about to apologize for confusing them when, as if by magic, a dog with a cropped tail walked down the road toward us. Both girls excitedly smiled and shouted 'Four and a half is the right answer, it's four and a half.' As they quickly walked away from me to pat the dog, I was so thankful that it had come to the rescue. It should have received a hero's medal.

The next day, when the school bus arrived, I noticed that the girls were met by their mother who drove them home, probably to protect them from me, some crazy guy who was harassing them with stupid mathematics questions.

This incident helped me recognize that the dualistic stress situation that I put the girls in is the same stress that faces each of us in industrial society from the day we are born. We learn to think habitually and mathematically, act and relate to each other and the world as if a dog has five legs. We are socialized to "as if" thinking that is maligned. We, in turn, assault each other and the environment as we demand that our natural life performs in five-leg ways when we only have 4-legs.

The science of this phenomenon has long been clear. The picture in our mind/consciousness "screen" (Sense #43) is experienced and reacted to by our body and spirit no differently than an authentic real-time experience. For example, if you think about sucking a lemon, you begin to salivate.

It is no surprise, then, that most people answer 'five' when I ask the "5-legged dog" question. Our mind/consciousness is indoctrinated or addicted to this answer by rewards and punishments concerning mathematical ability in our Industrial Society education, training and conditioning.

However, there is another reality at work. It's the nature of our planet, Earth, and the solar system. Its eons of natural grace, self-correcting intelligence, and restorative attraction typically produces dogs with four legs, dogs with tails that are not legs. Dogs know this as part of the dogness intelligence.

Nature's authentic reality also produces the way the world works as an optimum of life in diversity and balance, as a web of life with a perfection of its own that does not produce the pollution and disorders that result from our "tail is a leg" thinking.

We are born into and are part of nature's reality. When we choose to let it touch us, it gives us the common sense to recognize that a warped, but habitual, "tail is a leg" way of thinking is wrong concerning nature's organic perfection by which it works and that we inherit as part of nature. It is the self-evident purity and happiness of life in balance. That wrong underlies the "unsolvable" personal and global troubles that we suffer in Industrial Society. In this regard, our thinking is no better than a rabbit doing the bookkeeping for Wal-Mart. (See many examples at http://www.ecopsych.com/nineleg.html)

Where is your life in this story? How does it affect your past and present, your dreams and future? *(Grok this now. What does nature "say"?)*

Story #8. The Respected Island

In the State of Washington, the USA just off the shore of San Juan Island National Park at its English Camp is Guss Island. It consists of a few acres that are held sacred by the Lummi Native American tribe as the place that they entered the world. The island is respected and protected as such by the U.S. National Park Service, no visitors are allowed. Nature holds the wooded island together with a peaceful integrity that includes humanity protecting it. Nature's unifying "attraction glue" sustains the unadulterated virtue of the island and people while the rest of the world is increasingly falling apart.

Guss Island is an extraordinary organic statement. Its purity proclaims that part of the mentality of contemporary society reveres and can relate to the natural life of our Planet. Its wholeness displays the wisdom of Nature's balance and beauty in comparison to modern society's excessive exploitation of the natural world accompanied by its deterioration of places and people.

The island demonstrates what we have long known: Nature consists of a wondrous, self-correcting "glue" that unifies things in pure and cooperative ways that do not produce garbage. It does the same for the life of our Planet Earth island as it does for Guss Island.

Unification is the solution for most problems. It works if you apply the natural world's "gluing power" to relationships to unify them.

The Testimony of Steven Rector

A Personal GreenWave-54 Field Journal Entry (with numbered links and references to additional information.)

> "While I was in a protected natural area near Guss Island, Washington, learning how to use GreenWave-54 ecopsychology, David, our mentor, introduced the four of us to it by involving us in its warrantied process (senses #18, #7). Our exploration by experiencing it firsthand discovered that:
>
> - Whatever our senses experience in Nature is a self-evident fact for us at that moment, a fact that we do not need to defend because self-evidence is undeniable (sense #18).
> - We could reasonably see that we had many more than five senses operating because we experienced, spoke and identified some of them: hearing, distance, gravity, self, language, consciousness, aliveness, humor, contrast and reason (sense #6).
> - We could validate that nowhere here or anywhere else could we find any evidence of Nature communicating through stories or labels. This was an attribute of humanity alone (sense #18).
> - To help us know the life of Nature as it knows its nameless and speechless self, we spent five minutes quietly in the woodland identifying each thing that we became aware of there as "nameless" or "story-less." Attractive meaning and value were added to many ideas we held about the natural world. We then got together and shared what we sensed and felt from doing this (sense #22). We discovered that by dropping our labels and stories what one person said they had experienced each of us sensed as well since we were still unified in and by the forest. This made sense to us because everything is a supportive part of Nature except our nature-disconnecting stories (sense #17).

We helped each other realize that our love for or attraction to Nature that we were exploring was our 54 natural senses[27] organically registering Albert Einstein's Higgs Boson Unified Attraction Field attracting all things into consciously belonging in the Universe's time and space of the moment (senses #1, #11, #15,). Then we soloed for the next five minutes while labeling everything we became aware of, including ourselves, as 'Unified Field Attraction Love' (sense #2).

When we came together again as a group, a stronger unifying feeling was undeniable. Now our senses of love, trust, place, community and time became apparent as our forest community experience validated them. Folks spontaneously shared how this same feeling of well-being had helped them and others when they were ill or suffered disorders (sense #2).

We noted that some folks addict to artificially producing this supportive feeling by using drugs, alcohol or excessively dependent relationships. These detached them, short-term, from their upsetting nature-disconnecting stories and gave them relief or emotional rewards (3). However, these satisfactions were accompanied by harmful after effects. Applying Organic Psychology could replace this short circuit by supporting the health of the natural world in and around us continually. (senses #4, #5).

The things we discovered from this experience were

- How amazingly diverse Nature is,
- How we loved being aware of and in Nature,
- How each element in a natural area was a unique and attractive individual, including each of us,
- How it felt right that everything was right there to experience and love in the moment,
- That we felt relieved by not having to label things "correctly" or at all,
- That we found many wonderful new things about Nature by removing stories and labels from them and that this made us feel closer to them.
- That a "brightening" or vibrancy of things took place after awhile when we called them "Nameless." We could hear things we didn't hear moments earlier,
- Feeling a greater belongingness to everything including each other when we called ourselves "nameless,"
- Our customary meditation process benefited from a new unifying dimension, calling human-built structures and effects "blueprints" made us feel more reasonably able to control them.

We validated that these discoveries were real for each of us because we experienced them, they were self-evident, they registered directly on our senses (sense #19).

We walked back to the beginning of the trail labeling things we experienced as "nameless" "love," "attraction" or "Unified Field" and felt a greater wholeness than when we started. Then, David, had us pinch ourselves until it hurt so much that we stopped. We explored how our sense of pain was not a negative instead it was an attraction. It was Nature's protective attraction that signaled us to find more satisfying and reasonable attractions. We recognized that our sensations of anxiety, depression, and anguish, our senses 25-27, also make this contribution to our welfare (sense #6) and we could achieve this at will through nature-connecting activities.

We validated that moment-by-moment everything was attractively connected and as one as part of each moment of the Universe's Unified Field (senses #2, #23). When we consciously thought that plants or we were alive, the Earth and Universe also had "to be," alive and flourish for us because everything was an equal and identical Unified Field attraction at that moment including what we sensed and felt (sense #16).

We ended up looking at the aliveness of clouds as they moved across the sky into beautiful new shapes and we felt harmony and peace knowing we were doing the same thing with them and each other, no matter our cultural or genetic differences (sense #9). We noted that people in the middle of a city could do this with clouds, parks, and weeds (sense #14). Then David distributed sheets with the 54-senses listed on them and activities we could do to strengthen what we had just discovered (get one here[47], sense #13).

What fascinated me was that using GreenWave-54 we learned all this through trustable experiences in Nature, the real thing, in less than an hour. This was because what we were learning we could sense and feel right there around and in us immediately. This was powerfully different from the isolation generated by words in a book or lecture, words that Nature could not even register no less consider.

GreenWave-54 was enabling us to be whole life reality, not just to abstract it with stories right or wrong (sense #21). We were sensing and feeling that we were helping our 54 natural senses remember what they already knew. It felt attractive to give them safe time and space to connect with themselves in a natural area.

We concluded by validating that in Nature what we were attracted to was doing the attracting in a balanced way. I recognized that at my school it would take a full year science and philosophy course to get the same results if this was even possible (sense #10). Could an indoor course ever substitute for learning how Nature works from authentic Nature, the fountainhead of authority in how it works? (sense #20).

"The only source of knowledge is experience." – Albert Einstein

Where is your life in this story? How does it affect your past and present, your dreams and future? *(Grok this now. What does nature "say"?)*

Expedition Experiences that Actualize the Stories

---

47 http://www.ecopsych.com/MAKESENSEWALK.docx

As mentioned, this book's expedition into the essence of attraction in natural areas, backyard or backcountry, explores our central culture's space/time relationships. This is like knowing the Sun has risen and is shining because we experience it yet not knowing if it will rise and shine tomorrow.

Revolutionary Wisdom activates our 54 natural attraction senses[27] to help us address the source of today's nature-prejudiced catastrophe. The latter excessively detaches, on average, over 98 percent of our senses, time and thinking from the beneficial, self-correcting ways of the life of Nature.

Industrial Society makes us suffer "Earth Misery" a runaway, destructive, natural resource, species extinction and mental illness calamity[5]. It deteriorates personal, social and environmental well-being by ignoring the natural world's GreenWave-54 wisdom.

Without telling stories or producing garbage, moment-by-moment the organic life of Nature attractively creates Earth's pure optimums of life, diversity, cooperation, balance, beauty, and well-being. Our human being contains all of them.

Five decades of all-season Organic Psychology research into the restorative essence of natural areas have validated the therapeutic benefits of engaging in a GreenWave Unified Field process. It strengthens any endeavor by scientifically substantiating sensations and feelings as self-evident facts of life that help us enjoy the happiness of whole-life accuracy. The art of its hands-on, core learning process empowers us to create sensory natural area moments that let Earth teach and heal.

Story-estranged from Nature's unifying ways that they inherit, most people and institutions are at odds or war with the nature of each other and the life of planet because they are convinced to overlook that our 54 natural attraction senses are:

- REASONABLE: Mathematics, Archimedes' abstract "Mechanical Theorems," is considered to be purely rational fact. However, the self-evident truth that humanity inherently senses feels and learns from natural attractions is more ancient, real and universal. For example:

    The sense of taste: when a natural substance is sweet it is attractive, often edible and digestible, it rewardingly satisfies our sense of hunger so, in time, our sense of reason may validate the substance as food and allow us to eat it again.

The sense of temperature: If the heat from a fire burns a person, they may learn to trust that sensation is attracting them to use their sense of reason or fear to find more attractive things or attractions.

- RELIABLE: Our senses and sensations register immediately and pure in us as part of the now of our time/space Universe. The passing of time and influence of stories seldom adulterate them.
- AVAILABLE: For survival, nature endows humanity to continually register attraction sensations and feelings on conscious and subconscious levels.
- TRUSTABLE: For example, if people pinch themselves they register and trust that they feel something. They believe they will feel something again if they pinch themselves again.
- REPEATABLE: For example, if people pinch themselves many times over a period of time, they feel something every time. Other people demonstrate the same attribute.
- GLOBAL: For example, Humanity throughout the world senses and feels as well as trusts that the ability to sense and feel exists in themselves and others.
- TIMELESS: Records show, and we commonly reason that humanity, the past, present, and future contain, and will contain, the ability to sense and feel.
- PREDICTABLE: No matter where humanity goes if people are conscious they will sense or feel attractions. Consciousness itself is an attractive sensation or feeling.
- DIVERSE: There are at least 54 distinct natural attraction sensations and feelings humanity can register and thereby know the world (sense #26).
- EDUCATIONAL: Sensations and feelings provide humanity with a wide range of survival information.
- VALUABLE: Sensations and feelings enhance survival potentials as well as help establish a strong sense of self. "I feel therefore I am," is as true, significant and important as, "I think therefore I am."
- HONEST: Sensations and feelings always offer us excellent information about the state of our being.
- WIDE RANGED: Sensations and feelings help humanity register the many attraction sensitivities displayed throughout nature by the plant, animal and mineral kingdoms.
- DEMONSTRABLE: Humanity can often register what other members of humanity are sensing and feeling.
- INDEPENDENT; Humanity can register and reason with attraction sensations and feeling that lie outside the current or destructive operant and dogmas of their culture or society.

- SPIRITUAL: Natural sensations and feelings enable people to register and relate to nature connected aspects of spirit and soul that, to our loss, some parts of society omit.
- ATTRACTIVE: Humanity embraces and seeks sensation, it is attractive, we usually never desire to give up our ability to sense and feel.
- INTELLIGENT: Our natural senses are the intelligence of the life of our Planet as exemplified by the sense of Thirst that innately knows water is available, is attracted to turn on when we need it and to turn off when we have enough. Although each of our 54-senses is part of our sense of reason's ability to make intelligent decisions, we are brainwashed to believe we only have five senses yet we know that thirst is not one of them.
- FREE: Sensations and feelings are of, by and from nature, no culture or individual lays claim to inventing or owning them or legally restricting their availability.
- SELF-REGULATING: Sensations and feelings help regulate and guide each other homeostatically. For example, if a person is attracted to pinch themselves too hard, another sense or senses (sense of pain, sense of reason, or both) attract them to find more attractive things or behaviors.

What happens to the life of Earth happens to us for in any given moment the essence of our lives is identical. Our stories excessively socialize and pay us to disconnect from, compete with or subjugate and exploit the self-correcting life of Earth that ordinarily supports all of life in good health, including us. This disconnection and abuse make us suffer the misery of our wide range of conflicts, disorders, emotional wounds and discontents. The inborn peace and love of our lives are socialized to produce and endure the injuries of relationships like those in of competing football teams rather than cooperatively build mutually supportive relationships with the vulnerable, non-story life of Earth. Today the score is 4900 to 4500 nuclear warheads with Russia ahead of the USA.

In the past fifty years, our undeclared war on the life of our planet has increasingly resulted in the excessive stress and corruption; we endure in contemporary life. It has produced a catastrophic forty-five percent reduction in species, natural resources, and mental health[5]. At this moment, this desensitizing calamity continues to grow as our education and therapies leave us gasping because they support us gaining more money and prestige from our "normal" corruption, dysfunction, environmental degradation, waste, disenchantment, and inequality.

Education and therapy today seldom teach us at the start that we agonize our problems because the resources and wellness of our planet that generally would support us are severely diminished and are further diminishing.

Revolutionary wisdom is appalled that our Ego denies that the way it thinks and relates is deteriorating the life and sanity of Earth. Wisdom sees this as being suicide by denial.

Most people are so alarmed by our earth-misery debacle that they deaden themselves to it. Equally as frightening is that nobody shows us the prime cause of this fiasco or provides us with remedies for it. **Our experts identify the problems that face us. However, we increasingly suffer our disorders because they do not give us the tools to correct our core conflicts.** The shrug of their shoulders is not sufficient.

Without the tools we need, we erode morality and human decency as we continue to torture, child abuse[48], racism, emotional disorders, sexism, homophobia, violence and anti-Semitism adnauseum. These result from our environmental and human injustice that is addicted to excessively conquer the sane way of life of Nature, in, around and as us, a crime that gives Nature no legal rights.

What we do to the life of our planet, we do to ourselves because its life is our life. Our society's "artificial indoor closet lifestyle" that we have built to isolate us from Nature's fluctuations began to overuse the natural world as a closet-building resource in 1974. This has increasingly deteriorated our personal, social and environmental relationships since then. The closet is now vulnerable to Nature's response to its overuse and our increasing problems result. Those who deny this whole life fact help continue this catastrophe. For example, all the economic "growth" we experience or plan to make neglects to mention that it is a corrupt madness because we are half a planet bankrupt and counting. We are already paying the piper with decreased well-being on most levels.

> **"You see, nature will do exactly what it must and if we are a hindrance to its development, to even its destructive powers to reform itself and we are in a way, we will go." - Ralph Steadman**

Where is your life with regard to the information, above? How does it affect your past and present, your dreams and future? *(Grok this now. What does nature "say"?*

EXPLORE NOW: VALIDATE. *CONNECT THE CONCEPT, ABOVE, WITH A CONSENTING NATURAL AREA.*
• List the attractions and senses[40] that you find or that find you. (Appendix A: Our Fifty-Four Natural Senses and Sensitivities)

[48] http://www.ecopsych.com/journalproposal.html

• Add this process to an experience you have had in your special area of interest: Art, Creative Writing, Music, Yoga, Parenting, Recovery, Addiction, Renewal, etc.
• **Supportive Reading and Activities:** in a natural area read the next chapter in *Reconnecting with Nature* (or approved course book) including doing its activities and journaling them.
• **Journal Response Form:** For personal growth and reference to describe the value of this section, please complete the Appendix B: Revolutionary Wisdom Response Form: Journal Response Form[41].

#2A. An Expedition Challenge

Revolutionary Wisdom's art and science of the Organic Psychology GreenWave-54 Unified Field (GreenWave-54) provides the details that we need to come into balance therapeutically.

- It adds a universal mathematics of attraction/love to any mentality, relationship, sensation or endeavor and this makes them organic. It enables them to plug into the self-correcting ways that Nature works.
- It equips us with scientific methods and materials that help us be part of how Earth's brilliance operates rather than be injuriously disconnected from its homeostatic well-being and suffer accordingly.

Be assured that these tools are available. The pages of *Revolutionary Wisdom* are them; you have them in hand. By teaching us how to use its process, on personal and global levels *Revolutionary Wisdom* helps us identify and address the specific source of our self-destructive and nature-deteriorating ways that we are attached or addicted to, methods that numb us so we cannot quickly change. It also recognizes that it is impossible for us to stop our undeclared war with nature in and around us without simultaneously obtaining nature-connected therapy and happiness to counteract that war's crippling effects. The fact that the war has crippled us prevents us from stopping it.

We fail to realize that on our nature-ailing planet, the nature-estranged therapies and thinking that we presently use in our nature-crushing society, in the long run, help us more robustly increase our ruinous manipulation of Earth's life. For this reason, we increasingly suffer the consequences. *(Grok this. Enjoy the benefits of Organic Psychology so you can teach others to do the same.)*

Most of our disorders result from us being influenced to use the runaway science and technology that causes them and not to use the wisdom of organic science and technology that helps us correct them.

EXPLORE NOW: VALIDATE. *CONNECT THE CONCEPT, ABOVE, WITH A CONSENTING NATURAL AREA.*
• List the attractions and senses[40] that you find or that find you. (Appendix A: Our Fifty-Four Natural Senses and Sensitivities) • Add this process to an experience you have had in your special area of interest: Art, Creative Writing, Music, Yoga, Parenting, Recovery, Addiction, Renewal, etc.
• **Supportive Reading and Activities:** in a natural area read the next chapter in *Reconnecting with Nature* (or approved course book) including doing its activities and journaling them.
• **Journal Response Form:** For personal growth and reference to describe the value of this section, please complete the Appendix B: Revolutionary Wisdom Response Form: Journal Response Form[41].

#3 An Expedition Challenge

⇒ Identify the most reliable attraction or love that you have in a natural area or thing, (pet, plant, etc.) and then Grok it.

⇒ Thank its life for contributing to your life.

⇒ Imagine that somebody's story or a law stopped you from having this relationship and removed this love of yours from you.

⇒ How would you feel? What would you do?

⇒ How is this different than the natural life of our psyche being excessively separated from the natural life of Planet Earth especially since the two are identical? *(Grok this. Enjoy the benefits of Organic Psychology so you can teach others to do the same.)*

"What we call wildness is a civilization other than our own." Henry David Thoreau

A potential final exam question: *Why do we continue to produce and suffer our unsolvable problems when we have already identified them?* (*who, what, where, how, when*)

- **Practice educating, counseling and healing with Nature**

  Except for excessively industrialized humanity, the life of Earth's animals, plants, minerals, and energies are unified by attraction to each other as part of their attraction to survive as the life of Earth and its survival. They cooperatively support each other as if married for eternity. Within humanity, this same attraction is registered by at least 54 natural senses that we as well as the web-of-life *hold in common.* Also, the Web has many additional attractions operating that we do not register but whose processes and results we can tap into.

  You may find it beneficial now to read about and identify our 54 natural attraction senses that we share with Nature. Explore which of these senses you have already experienced in a natural area and with people. (See Appendix A: Our Fifty-Four Natural Senses and Sensitivities)

If you want to begin immediately working with them online with others, you are encouraged to take our course[29] in this skill. You can get credit for it if that will be helpful to you. You can petition for financial assistance if needed.

From sub-atomics to weather systems and beyond, in scientifically valid, organic purity, balance and beauty Earth's unadulterated web-of-life operates so that everything belongs. A profound love of space/time whole of life is shared by and with the life of each part of the web simultaneously.

In this book, Project NatureConnect's reviewed and accredited Organic Psychology helps your senses interlace with the joy of Nature's integrity and wellness that are described in Michael J. Cohen's books:

- *Reconnecting with Nature,*
- *Educating, Counseling and Healing with Nature,*
- *and With Justice for All*

Revolutionary Wisdom is an empowering whole-life art and science that works best when you use it to help others benefit from using it.

Additional Information: http://www.ecopsych.com/mjcohen.html

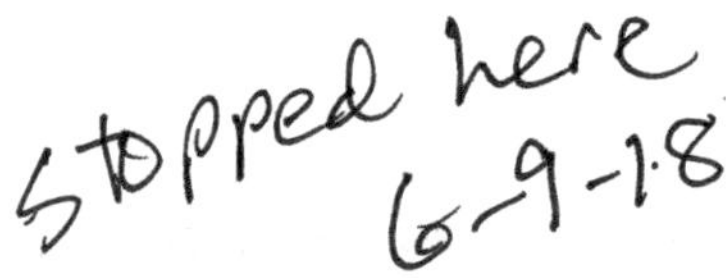

#4 An Expedition Challenge:

Identify and name which of our 54-senses you have already experienced in a natural area and with people. (See Appendix A: Our Fifty-Four Natural Senses and Sensitivities)

A Potential final exam question: *Why do we continue to produce and suffer problems when we know what they are and that we cause them?*

Learning and Relating Through Natural Attractions

By engaging in the same expedition education process (see CHAPTER ELEVEN: An Expedition Process Summary) that created the ways and means of Organic Psychology, *Revolutionary Wisdom* helps its readers produce reasonable sensory relationships with the natural world, in and about us, as they learn to apply Nature's intelligence. As mentioned earlier, in the expedition challenge, above, *this continues here and now by further contacting the natural world as part of this book's learning, growth and development process* rather than omitting it as is all too common.

From this point on, through these pages, the whole of the natural world itself directly contributes its powers and truths to our benefit. Its reality becomes part our text and resource library so we can learn about it by doing as well as practice what we learn and teach others to do the same. **The secret to doing this is to learn how to habitually use Nature itself as a library to obtain accurate information about it.** Most of this information is already in you as part of nature. You just need to learn how to locate and contact it.

Because authentic Nature is the fountainhead of authority in how it works, every natural area displays and exemplifies how Nature produces its self-correcting balance and beauty. This is because that's what Nature knows how to do best and it is doing it there. **To repeat,** for you to beneficially connect with its wisdom and energies *only continue to read this book while you are consciously in contact with the most attractive natural area or thing that attracts you, backyard or backcountry.* The more natural the area, the better. A weed, insect, tree, pet or the sky and clouds are acceptable when backyards, parks and other natural areas are not available.

In time contact can be made with the genuine nature of another person, too, once its essence is identified with them.
Once you learn how to create moments that connect with the life of Nature and let Earth speak, use them to help your education and life experience validate these two facts

1. On average, in contemporary society, over 95 percent of our time and 99 percent our thinking and feeling are excessively disconnected from the self-correcting ways that the life of authentic Nature work in and around us.

Instead, we spend this time in homes, schools, vehicles, and business or entertainment buildings while communicating and learning through **stories and media that the life of Nature does not understand or use because it is silent, it does not speak, read or write.** Our sense of literacy is a unique human quality for our survival in equilibrium with the web-of-life ***when we choose to use it that way.***

2. The life of Nature and Planet Earth organically produce their optimum balance of life, diversity, beauty, purity, peace, wellness and cooperation without producing our runaway garbage, stress, and abusiveness.

The excessive separation of our psyche from Mother Earth's organic life and then bonding it to nature-substitutes that have destructive side effects is a global emotional abortion that produces our feelings of alienation and wanting. They, in turn, motivate our unreasonable excessiveness because we sense and feel that there is never enough.

To think the above is not true is like having a vasectomy because you do not want to have any children, then going home and there they are.

Additional Information: http://www.ecopsych.com/referencearticle.html

"Death is not the greatest loss in life. The greatest loss is what dies inside us while we live." - Norman Cousins

Where is your life with regard to the information, above? How does it affect your past and present, your dreams and future? *(Grok this. Enjoy the benefits of Organic Psychology so you can teach others to do the same.)*

A potential final exam question: *Why is it important to go to the fountainhead of knowledge about how the whole of life produces its balance and beauty? (who, what, where, how, when)*

EXPLORE NOW: VALIDATE. *CONNECT THE CONCEPT, ABOVE, WITH A CONSENTING NATURAL AREA.*
• List the attractions and senses[40] that you find or that find you. (Appendix A: Our Fifty-Four Natural Senses and Sensitivities)

• Add this process to an experience you have had in your special area of interest: Art, Creative Writing, Music, Yoga, Parenting, Recovery, Addiction, Renewal, etc.
• **Supportive Reading and Activities:** in a natural area read the next chapter in *Reconnecting with Nature* (or approved course book) including doing its activities and journaling them.
• **Journal Response Form:** For personal growth and reference to describe the value of this section, please complete the Appendix B: Revolutionary Wisdom Response Form: Journal Response Form[41].

PART FOUR

Grokking the Essence: Practical How, What and Why Activities

"I have been training an 8-day coaching class divided into two modules of 4 days each with 2 days off in between. I usually don't work with big groups, and this group was a group of 16 people from different places in Argentina and Latin America. In this study group, you all know my personal changes in the way I feel and how my state has been really stable and centered and happy since doing these activities, and when I felt off, I would go back to nature to reconnect.

It is hard to put this into words… but working with this coaching class and being myself in a different way of thinking, and connecting with the beauty, potential and connecting with them from a completely different place, maybe it was a literate transaction, but I felt this same feeling I get when I connect with nature, and they did too. I had CEOs, managers, consultant's lawyers, etc., who started the training really stiff and they told me by the end they were other people… that I had helped them reconnect with their true nature, their passion for life and with the beautiful things in life. The material was the same, but I was different. They had the same feeling I had while connecting with the earth. They felt respected in their own individuality and contribution, and we all really connected to each other. In this way, we restored the attraction strands that connected us as people.

The class has been the best group and experience I have ever had as a trainer in the last 15 years. They all took care of each other's well-being, and it was beautiful to see them blooming, kind of shining and finding their way back home to who they really are and want to be, forgetting about their pre-conceived stories. We all celebrated this sacred space!

It is so hard to put this experience into words (it is very emotional for me, and my smile is soooo big!!!). What we lived was connection, love, beauty, respect and real LIFE! Now I realize how Organic Psychology not only has helped me reconnect but also how it enriches and makes my relationships healthier and more transcending."

CHAPTER ONE: Our Story World

This scenario, repeated from the Introduction, describes our most significant challenge today and how to remedy it:

Industrial Society has emotionally attached us to drive an advanced technology automobile that, because it is not entirely organic, it makes us produce and suffer our local and global misery [5]. This is because to satisfy our practical needs and emotional frustrations we enjoy the fulfillment of our sense of motion speeding our car down the highway.

Suddenly we see that we are going to crash into a group of families having a picnic in a beautiful natural area.

Because we have not yet learned how to fully activate the car's advanced braking and steering system, in anguish, we hope and pray that it will change direction as we fearfully scream "Oh my god" "stop" or "whoa," as if the vehicle was a runaway spirit or horse.

These reactions are unscientific and outdated. They do not stop our advanced technology automobile. For this reason, we m/m wreak havoc on innocent people, places and things including ourselves as passengers.

What Industrial Society's science knows but seldom teaches us is that, as in real life, every atom, energy, and relationship in this scenario was once part of the life of a natural area of Earth. There, in congress with other atoms and energies, and without using stories, they organized themselves and related to balanced, self-correcting diversity, purity, and growth. This is significant because of **these same organically sound relationships are right now operating in the natural areas that our runaway car is tearing asunder.**

On many levels, we sense and are emotionally hurt by the uncontrolled car injuring us because **the life of Nature in and around us is identical.** Our story that we are special and immune to or protected from the car's destructiveness is a self-inflicted delusion.

> The global misery[5] all too many of us have experienced is like us being a kitten that was harnessed to the bumper of a tractor-trailer truck to pull it out of a ditch. Somebody told the driver the kitten could not do that. "Why not? I've got a whip," he replied, "or I could raise their pay."

The art and science of Organic Psychology empower us to learn how to let the life-wisdom of any natural area teach us to m/m transform the misery from our runaway automobile into peace, health, and balance that stops the vehicle. Its 54-sense process helps us safely Grok the ways of Nature and let them bring out within us their homeostatic purification powers that have beautifully supported the life of Earth, including us, develop and grow its stable integrity over the eons. This centering and healing procedure reasonably enables our story way of knowing to energize the subdued natural world's qualities in us so we may consciously register, enjoy and apply them.

> *We need mental repair when we believe that we can replace our car's broken computer with an old horseshoe and our self-evident, whole life experiences do not change this belief.*

Because we are an advanced science and technology-based society, the scenario, above, makes it obvious as to what we scientifically need to do to reduce the anguish we are causing our planet and ourselves. ***We must stop merely reading and telling this story, here and now.*** This is because the remedy for the scenario is not the story. It is to scientifically learn how to m/m create good relationships with Nature by establishing them as we Grok authentic nature and the latter is not on this page. It is in natural areas, and they do not speak, use or understand stories. For this reason

> We must visit a real natural area accompanied by this book and its activities. Doing this enables us to create Grokking moments in "wildness" that let Nature touch and energize its sensory, self-correcting ways within us. They are a seamless continuum of it and its homeostatic community wisdom of the eons that is also our natural body mind and spirit.

> When we visit a natural area, our 54 natural senses are in m/m loving contact with their origins and source in the life of nature. Moment by moment, they know what to do to survive sensibly, and they do it. The feeling-wonderful effects of good experiences that most people have had in a natural area confirm the therapeutic value of this happy connection of our senses with Nature's unseen intelligence. It is contacted with the real life of Earth, not communicate with an artificial escape from reality.

The above explains why the first instructions for using this book **are only to read it and Grok its paragraphs while you are m/m in a natural area or conscious contact with a genuine part of the web-of-life**, not just a poem, film or recording of it. The latter are some of the thousands of media substitutes we use for authentic natural area contact. We suffer because when it comes to the essence of Nature's life and its eons of wisdom, **there is no substitute for the real thing.** The tools in this educating, counseling and healing with Nature book empower you to plug yourself into the m/m, self-correcting, whole-life reality core of Nature that Industrial Society excessively makes us overlook and subdue.

Where is your life with regard to the information, above? How does it affect your past and present, your dreams and future?

Additional Information: http://www.ecopsych.com/thesisquote6.html

> **A Self-Evident Fact:** Nature and Earth neither use nor understand stories.
> **A Supportive Observation:** *Show me one place where Nature uses stories, and I'll show you the one and only Bugs Bunny.*

A Potential final exam question: *Why do new reasonable stories seldom change our behavior about how we excessively relate to nature destructively?*

EXPLORE NOW: VALIDATE. *CONNECT THE CONCEPT, ABOVE, WITH A CONSENTING NATURAL AREA.*
• List the attractions and senses[40] that you find or that find you. (Appendix A: Our Fifty-Four Natural Senses and Sensitivities) • Add this process to an experience you have had in your special area of interest: Art, Creative Writing, Music, Yoga, Parenting, Recovery, Addiction, Renewal, etc.
• **Supportive Reading and Activities:** in a natural area read the next chapter in *Reconnecting with Nature* (or approved course book) including doing its activities and journaling them.

- **Journal Response Form:** For personal growth and reference to describe the value of this section, please complete the Appendix B: Revolutionary Wisdom Response Form: Journal Response Form[41].

#5 Explore Nature Challenge:

- Visit a natural area near a building or machine.
- Identify what happens in the area with how it is artificially replaced by what happens in the building.

A Potential final exam question: *Where in the solar system do you live?*

EXPLORE NOW: VALIDATE. *CONNECT THE CONCEPT, ABOVE, WITH A CONSENTING NATURAL AREA.*
• List the attractions and senses[40] that you find or that find you. (Appendix A: Our Fifty-Four Natural Senses and Sensitivities) • Add this process to an experience you have had in your special area of interest: Art, Creative Writing, Music, Yoga, Parenting, Recovery, Addiction, Renewal, etc.
• **Supportive Reading and Activities:** in a natural area read the next chapter in *Reconnecting with Nature* (or approved course book) including doing its activities and journaling them.
• **Journal Response Form:** For personal growth and reference to describe the value of this section, please complete the Appendix B: Revolutionary Wisdom Response Form: Journal Response Form[41].

THE WHOLE TRUTH

As the runaway car scenario that begins this chapter shows, this book starts from the fact that most of us do not pay full attention to the established, evidence-based, scientific knowledge and ways of Industrial Society that wisely exclude the unprovable. These m/m truths include that

- We mostly communicate and build relationships through words and stories while Nature does not have this ability. This is a significant difference and challenge.

- We trust, depend on and are sustained by our central Big Bang science and technology stories that confirm that the Universe makes its own time and space moment-by-moment; we only exist in each immediate moment.
- Our science, technology, and profits are out-of-control attractive to us.
- The life of Nature organizes and produces its self-correcting balance, purity, and beauty.
- Our attachments to our inaccurate stories make us build and experience personal and global miseries that we cannot stop without help.
- The attraction-based aliveness of Nature does not produce our troubles. They did not exist before our nature-conquering stories came into play. **This fact is demonstrated by the difference in the well-being of the web of life in North America before and after the arrival of Columbus.**
- The aliveness of every seed, egg, and mineral is attracted to grow and support the balance and beauty of Nature's life without using abstract stories. **The opposite of attraction (to pull together) is an abstraction (to pull apart).** In all-attraction Nature: Together = Organic.

  Apart = Non-Organic Artifact
- Our indoctrination teaches us that we. know and learn about the world through five senses: touch, taste, sight, sound, and smell. This leads us to omit the life of our equally true 49 additional natural attraction senses[27] including thirst, community, reason, consciousness, music, balance and gravity. **Note that only one of the latter is one of the five.**

Additional Information:
http://www.ecopsych.com/truthlist.html
http://www.ecopsych.com/coretruth.html

#5 An Expedition Challenge

Visit a Natural area and find examples thereof each of the whole-life truths that are listed above.

A Potential final exam question: *What makes us undergo our miseries when we do not want to suffer them?*

EXPLORE NOW: VALIDATE. *CONNECT THE CONCEPT, ABOVE, WITH A CONSENTING NATURAL AREA.*
• List the attractions and senses[40] that you find or that find you. (Appendix A: Our Fifty-Four Natural Senses and Sensitivities) • Add this process to an experience you have had in your special area of interest: Art, Creative Writing, Music, Yoga, Parenting, Recovery, Addiction, Renewal, etc.
• **Supportive Reading and Activities:** in a natural area read the next chapter in *Reconnecting with Nature* (or approved course book) including doing its activities and journaling them.
• **Journal Response Form:** For personal growth and reference to describe the value of this section, please complete the Appendix B: Revolutionary Wisdom Response Form: Journal Response Form[41].

SELF-EVIDENT EXPERIENCE

We suffer because our conflicted thoughts, stories, leaders and institutions teach us to excessively separate from[22], conquer and deteriorate the intelligent, unified life of Nature in and around us. ***That foolishness stops right here.***

It stops because the organically sound personal and global solutions that this book provides are based on prominent, **self-evident experiences. That makes the information here undeniably true** because it registers directly in a time/space moment with and as our unified 54 natural senses including our sense of reason. When we include them in our relationship building, we can sense and feel the good effects of reasonable, consensual, sensory contact with the wisdom of a natural area.

If you experience that you can see these words right now and you **trust that fact is irrefutably self-evident,** this expedition will prove of great service to you and yours.

Many folks respond to the statement, above, by saying that it is OK, but they need to verify it by having other sources tell them it is a legitimate fact. However, it is also these other sources that are the source of our earth miseries because they teach us to devalue the truth of what we sense and feel.

> **"Truth is by nature self-evident. As soon as you remove the cobwebs of ignorance that surround it, it shines clear." - Mahatma Gandhi**

In congress in a natural area, our 54-senses detect and register the m/m world on mechanical, thermal and chemical levels that strengthen its and our well-being. This helps us make more sense as we learn to trust our sense of reason and our experiences in natural areas.

Do you recognize that this 54-sense process is vital, but missing, an element of our established doctrines, philosophies, institutions, processes, relationships, leaders and spirituality?

Sensing and validating the ***how, what, why, when, where and meaning*** of our loving 54-senses is a substantial, therapeutic benefit of Organic Psychology.

> **"The senses, being the explorers of the world, open the way to knowledge." - Maria Montessori**

We are destructively bewildered and thus react when we learn to be out of touch with reality that **we can't answer this simple question "Which one of our five senses is our self-evident sense of hunger or place or motion?"**

- Softly stroke your cheek and say: "Feeling is being."
- When you laugh to say: "This feeling is real."
- Pinch yourself and say: "I feel. Therefore I am."
- Tell yourself: "Feelings are experiences." "Senses, sensations and feelings are facts that make life worthwhile."
- What would your life experience be if you lost some or all your sensations and feelings?

It is self-evident that most of our society is attached to hurtful or abusive, nature disconnecting stories. Much of the direction of our personal life is guided by avoiding things that push our buttons, which trigger this hurt onto our screen of consciousness so that we feel it. **To recover from misguided humanmade stories, it is reasonable for us to Grok and attach to nature's prime, evidence-based connection story**. This makes sense because mysticism wielding the sword of technology hurts the life of nature in and around us and spells disaster.

"Loss is nothing else but change, and change is Nature's delight." -Marcus Aurelius

"A new idea comes suddenly and in a rather intuitive way, but intuition is nothing but the outcome of earlier intellectual experience."- Albert Einstein

Where is your life with regard to the information, above? How does it affect your past and present, your dreams and future?

A Self-Evident Fact: Self-evidence is undeniably true and accurate.
A Supportive Observation: *Show me self-evidence that is inaccurate, and I'll show you the genome of Baron Munchausen.*

A potential final exam question: *What makes a fact self-evident? (who, where, how, when)*

EXPLORE NOW: VALIDATE. *CONNECT THE CONCEPT, ABOVE, WITH A CONSENTING NATURAL AREA.*
• List the attractions and senses[40] that you find or that find you. (Appendix A: Our Fifty-Four Natural Senses and Sensitivities) • Add this process to an experience you have had in your special area of interest: Art, Creative Writing, Music, Yoga, Parenting, Recovery, Addiction, Renewal, etc.
• **Supportive Reading and Activities:** in a natural area read the next chapter in *Reconnecting with Nature* (or approved course book) including doing its activities and journaling them.
• **Journal Response Form:** For personal growth and reference to describe the value of this section, please complete the Appendix B: Revolutionary Wisdom Response Form: Journal Response Form[41].

#6 An Expedition Challenge

Do you deserve to experience good feelings because in Nature good feelings support ongoing survival? Will you increase trust and act to sustain feelings that are most reasonably attractive and satisfying?

To meet this challenge, in a natural area obtain consent to use **S-E-V-M-R-A-T-C-I** at will and where indicated in this text.
Reinforce the truth of your natural sense experiences. SEVMRATCI (pronounced Sev-mer-rat-see) is to **S**ense-**E**njoy-**V**alidate-**M**atch-**R**esonate-**A**ppreciate-**T**rust **C**elebrate and **I**ntegrate what your senses register in natural areas.

Use one of your 54-senses, *your sense of color, as an example here.* The color was part of space and time long before humanity, or you came into being. You are attracted to a color because for survival it calls or connects to the part of you that is and that needs this color now. That is what makes it attractive.

- **Sense:** Check out different parts of your surroundings and move to the point that, colorwise, seems most attractive.
- **Enjoy:** Spend one minute enjoying your chosen place's or thing's color. For example; Enjoy the color of an orange leaf. Recognize that you deserve to enjoy every aspect of life, including color because you are alive.
- **Validate:** Acknowledge to yourself that it is true you are someone who finds this place's color attractive and enjoys it because you experienced it.
- **Match:** Match this color with the same color that you might see or feel within you, your color memories, associations, and mood. Write down the parts of your personality that are this color. Assume a physical posture or motion or emotion that matches this colorful entity's shape and imitate this entity's motions.
- **Resonate:** Register the color through 53 other senses[27], for example, music #45: try to hum this color. (This whole course consists of just doing this one activity in a natural area continually since all the senses attractively resonate with each other except nature-disconnected stories.)
- **Appreciate and Honor:** Thank this color mood which has attracted you for having given your life so much color and feeling. Honor this entity with some physical act, gift or spoken words. Honor it for contributing its color sensations to the global life community, too.
- **Trust:** Trust the rationality and feelings you obtain from this SEVMRATCI experience. Trust the love of Nature you may discover. SEVMRATCI helps Nature express itself. Trust SEVMRATCI.
- **Celebrate:** Write a poem, haiku or statement or assume a posture or motion which you feel states your good feelings about this SEVMRATCI event. Hold your position or motions for at least one minute. As Rollo May says, "If you do not listen to your being you will have betrayed yourself."
- **Integrate**: Upon completing the SEVMRATCI sequence ask yourself "What would I sense and feel if my natural ability to register this attraction connection **was taken away from me?"**

Where is your life with regard to the information, above? How does it affect your past and present, your dreams and future?

A potential final exam question: *What does SEVMRATCI stand for and what contribution does it make? (who, where, how, when)*

<table>
<tr><td>EXPLORE NOW: VALIDATE. CONNECT THE CONCEPT, ABOVE, WITH A CONSENTING NATURAL AREA.</td></tr>
<tr><td><ul><li>SEVMRATCI the most interesting felt-sense attractions and senses that you find or that find you[40].(Appendix A: Our Fifty-Four Natural Senses and Sensitivities)</li><li>Add this process to an experience you have had in your special area of interest: Art, Creative Writing, Music, Yoga, Parenting, Recovery, Addiction, Renewal, etc.</li></ul></td></tr>
<tr><td><ul><li>Supportive Reading and Activities: in a natural area read the next chapter in Reconnecting with Nature (or approved course book) including doing its activities and journaling them.</li></ul></td></tr>
<tr><td><ul><li>Journal Response Form: For personal growth and reference to describe the value of this section, please complete the Appendix B: Revolutionary Wisdom Response Form: Journal Response Form[41].</li></ul></td></tr>
</table>

CHAPTER TWO: The Non-Language World

Aliveness

The blanketing convenient central means of Industrial society omit that we need to operate our runaway technology vehicles in organically sound ways. Connecting nature's self-correcting powers to these operating areas of society increase their contribution to peace, health, and sanity.

- Technology-dependent life
- Trusting self-evident experience
- Earth is speechless
- Science identifies truth
- Validating natural aliveness
- Activating 54 natural sense facts
- Einstein's m/m Unified Field is natural to love.
- Cooperation produces the fittest
- Attraction is the essence of life/love
- Now is most trustable

The remainder of Part 2 of this book helps you recognize the absence of nature's m/m core in these areas and correct this deficit by having you genuinely connect to natural areas and replace what is missing in them. Doing this improves your thinking, feeling, and relationships forever. **It will strengthen your ability to walk the talk that you discover and establish in this chapter.**

THE M/M LIFE OF OUR SPEECHLESS PLANET

The moment-by-moment (m/m) life of Nature and Earth consists of non-story, mute ways of relating. Nobody has observed any members of the web-of-life communicating with words. Whales can't write letters of protest or petitions to stop us from harming them.

In contrast, over 99 percent of our human life we think and relate through abstract stories about things, be the stories right or wrong, rather than through direct m/m, 54-sense experiences, and relationships with things.

Books, including this one, consist of words and Nature neither uses or understands them. Words can only accurately convey how Nature works if the words describe an attractive sensory contact with Nature. Therefore your natural area companion relationship is an m/m essence of Organic Psychology.

If somebody built a little green wagon and later you paint it orange, what color is it?

You have misled yourself if you think it is just orange. It is also green, and then you changed it to what it was not, but what you wanted it to be. That parallels what our stories do to Earth when they are not scientifically organic. They socialize us to relate to an orange-painted earth inappropriately. The result is earth misery [5].

We are born into and enjoy a unified m/m oneness with Nature. Scientifically, the life of Nature, Planet Earth and humanity is identical with the one exception already mentioned; Earth is speechless and non-literate. **What our stories do to or for Earth's life m/m we do to or for our life and vice versa.** The loving essence or whole of life that we share with Earth is what it has been m/m since the beginning of time.

#7 An Expedition Challenge: Story-Self Meet Your Earth-Self

Get to know the life of Earth in and around you as it knows itself.

1. In a consenting natural area, find something that attracts you there and asks it to tell you who and what **it is** without it using labels or names for anything.

2. Repeat the above by asking yourself to tell the natural attraction and yourself you who and

what **you are** without using labels or names for anything.

Repeat 2, above, while substituting your 54 natural sensitivities and sensibilities[27] that you share with the life of Earth to replace the labels and names you omitted.

It makes sense that since nobody has observed any members of the web-of-life communicating with words, our sense of language telling nature-disconnecting stories is a "foreign" overpowering trespasser of our vulnerable green planet. Too often this invader paints Earth orange without its consent and then treats it accordingly. This increases Earth misery.

> **"That which is truth or a reality that has happened is a fact or belief that is accepted as true." - Merriam-Webster Dictionary**

Expand your knowledge and use of self-evident experience by integrating the information at http://www.ecopsych.com/54nineleg.html Then continue below.

Additional Information: http://www.ecopsych.com/54journalotherbody.html

EXPLORE NOW: VALIDATE. *CONNECT THE CONCEPT, ABOVE, WITH A CONSENTING NATURAL AREA.*
• **SEVMRATCI** the most interesting felt-sense attractions and senses that you find or that find you[40].(Appendix A: Our Fifty-Four Natural Senses and Sensitivities) • Add this process to an experience you have had in your special area of interest: Art, Creative Writing, Music, Yoga, Parenting, Recovery, Addiction, Renewal, etc.
• **Supportive Reading and Activities:** in a natural area read the next chapter in *Reconnecting with Nature* (or approved course book) including doing its activities and journaling them.
• **Journal Response Form:** For personal growth and reference to describe the value of this section, please complete the Appendix B: Revolutionary Wisdom Response Form: Journal Response Form[41].

#8 An Expedition Challenge: The Intelligence of Speechless Earth

1. Pick up six similarly sized sticks (or rocks) and place them before you.
2. Shut your eyes, and then pick up one stick.

3. With your eyes remaining closed, mark the selected stick with a pen. Then feel the stick all over until you familiarize yourself with its shape, texture and other idiosyncrasies.

4. With your eyes still shut, return the stick to the pile. Mix up the pile of sticks, and then pick up the sticks one at a time and feel them until you believe you have found the stick you selected and marked.

5. Now open your eyes. **Did you select the stick that you marked?**

6. Thank the sticks for helping you validate your many senses at work here.

7. SEVMRATCI the most stimulating senses that you experience from doing this activity.

> **"Nothing we use or hear or touch can be expressed in words that equal what is given by the senses."- Hannah Arendt**

A Potential final exam question: *In Organic Psychology what is the significant difference between the life of Planet Earth and human life?*

EXPLORE NOW: VALIDATE. *CONNECT THE CONCEPT, ABOVE, WITH A CONSENTING NATURAL AREA.*
• **SEVMRATCI** the most interesting felt-sense attractions and senses that you find or that find you[40].(Appendix A: Our Fifty-Four Natural Senses and Sensitivities) • Add this process to an experience you have had in your special area of interest: Art, Creative Writing, Music, Yoga, Parenting, Recovery, Addiction, Renewal, etc.
• **Supportive Reading and Activities:** in a natural area read the next chapter in *Reconnecting with Nature* (or approved course book) including doing its activities and journaling them.
• **Journal Response Form:** For personal growth and reference to describe the value of this section, please complete the Appendix B: Revolutionary Wisdom Response Form: Journal Response Form[41].

CHAPTER THREE: The M/M Life of Science

We are in the midst of a life-deteriorating global crisis that demands an organically sound personal and universal solution.

Our society centers around the use of Science and Technology. This book's scientific accuracy is built on self-evident facts that our traditional thinking and relationships usually bypass or demean.

Since the cradle of Western Civilization in Greece, circa 600 B.C., the results of scientific methodology have worked and been trusted because they are based on evidence that connects humanity with authentic nature and disregards the unprovable in building relationships.

Pure science produces pure truths because it is evidence-based: A) it includes felt-sense and empirical facts that are indisputable and B) it omits mystical, supernatural and other suspicious phenomena.

Impure science contains inaccurate warps or short-circuits that produce problems that it cannot solve because attachments limit it to warped or short-circuited inaccuracies that are not evidence-based

We are in trouble when our high-tech vehicle is using a roadmap made by the Flintstones.

In what ways does your consenting natural area contact affirm the following observations?

The life of Nature is pure and sensible. It knows how to attractively organize itself to create diversity, balance, cooperation, beauty, and wellness without producing garbage, pollution or abuse. (senses #25 - #27)

Nature loves us into being born with the natural attributes, above, along with the ability to create stories that consciously help us a map, strengthen and support life (or deteriorate it).

Do our story-built deterioration of Earth's life and the misery it produces make sense to you? Is it scientifically sensible?

Expand your knowledge about Earth Misery by considering http:www.ecopsych.com/zombie2.html

Additional Information: http://www.ecopsych.com/54testimonials.html

EXPLORE NOW: VALIDATE. *CONNECT THE CONCEPT, ABOVE, WITH A CONSENTING NATURAL AREA.*
• **SEVMRATCI** the most interesting felt-sense attractions and senses that you find or that find you[40].(Appendix A: Our Fifty-Four Natural Senses and Sensitivities) • Add this process to an experience you have had in your special area of interest: Art, Creative Writing, Music, Yoga, Parenting, Recovery, Addiction, Renewal, etc.

• **Supportive Reading and Activities:** in a natural area read the next chapter in *Reconnecting with Nature* (or approved course book) including doing its activities and journaling them.
• **Journal Response Form:** For personal growth and reference to describe the value of this section, please complete the Appendix B: Revolutionary Wisdom Response Form: Journal Response Form[41].

#9 An Expedition Challenge

Astutely, one way that Nature transforms itself is through fire. Fire changes burnables into minerals, carbon dioxide and water vapor upon which other members of the natural world feed. Plants "eat" soil fertilized and re-mineralized by fire. They breathe carbon dioxide produced by the fire[49].

Clouds are produced by fire's water vapor which the web-of-life then "drinks."

A slow form of fire, called respiration (burning calories), produces your body heat (your temperature) and your life energy. Fire is you.

Challenge: Are you are feeling alive, energetic, mesmerized, warm and thankful for Nature's thoughtful gift of fire as a scientific fact?

A Potential final exam question: *Why is pure science trustable?*

EXPLORE NOW: VALIDATE. *CONNECT THE CONCEPT, ABOVE, WITH A CONSENTING NATURAL AREA.*
• **SEVMRATCI** the most interesting felt-sense attractions and senses that you find or that find you[40].(Appendix A: Our Fifty-Four Natural Senses and Sensitivities) • Add this process to an experience you have had in your special area of interest: Art, Creative Writing, Music, Yoga, Parenting, Recovery, Addiction, Renewal, etc.
• **Supportive Reading and Activities:** in a natural area read the next chapter in *Reconnecting with Nature* (or approved course book) including doing its activities and journaling them.

49 http://www.ecopsych.com/naturepath.html

- **Journal Response Form:** For personal growth and reference to describe the value of this section, please complete the Appendix B: Revolutionary Wisdom Response Form: Journal Response Form[41].

CHAPTER FOUR: Earth as Living Organism

PLANET EARTH: THE LIFE THAT WE LIVE IN AND AS, BUT NOT ON.

Scientifically, we are part of the life of Nature. As a unified team, the wondrous ways of the web-of-life simultaneously dance and stream around, through and in us, moment-by-moment. However, because our society is excessively disconnected from Nature's wisdom, our nature-estranged sci/tech thinking and advanced technology tend to injure Nature's ways as well as hide them from our awareness. **This essential but missing information makes us cause the web-of-life to deteriorate.**

Organic Psychology helps us address our bizarre disconnection from Earth's web-of-life. It scientifically reconnects us with it in a natural area through the sensibilities of our 54 natural senses and sensations, including our sense of reason.

These senses are each indisputable, self-evident facts of life that our engrained objective research methods must include as valid evidence. The 54-sense science of Organic Psychology helps us remedy and prevent our destructive short circuits.

A natural area gives anyone the opportunity to "Grok" any authentic thing there so that we may learn to think, feel and act in synchronicity with the perfection of the life of Nature. Societies that include people who help others make this happen don't produce our runaway problems.

Recognize that your natural area contact is being made through some or all your intact 54-senses as described below. This is important to note because most of the information in studies, books, and references are produced by excessively nature-disconnected scholars who believe we only have five senses and what we experience, sense, and feel is not scientifically valuable because it is subjective, it may not be able to be measured or eliminated.

> **Your aliveness is part of the whole aliveness of Earth. When the quality of Earth's life deteriorates or ceases, so does yours.**

Your aliveness is scientifically subjective, your natural senses and feelings along with nature do not count nor do they have legal standing.

Visit a natural area and sense its wholeness. It is self-evident there that you are engulfed in the life of Earth's atmosphere, biosphere, mineral, and energy body. **The fact that its web-of-life flows around and through you and that every few years becomes you, and you it, makes you part of Earth's life. You live in and as it.**

> **A Self-Evident Fact:** We live in, not on, the life of Planet Earth
> **A Supportive Observation:** *Show me someone who says we don't live in Planet Earth and I'll show you a "we" from another planet or E.T. from Spielberg, USA.*

The life right now that you experience being kept alive by the dance of Earth's life flow and vice-versa. For example, in your life dance, plants give you oxygen, and the carbon dioxide you exhale helps keep plants alive[49]. You are life partners with them and the rest of the self-sustaining web-of-life, including sunlight.

Moment by moment, the life of the river and its flow is never the same. All you need do to begin benefiting from GreenWave-54 Organic Psychology is validate that you are alive. If you are reading these words, aren't you? Isn't your sense of aliveness a self-evident fact that you can trust and depend on right now?

> **"Do not dwell in the past, do not dream of the future, concentrate the mind on the present moment." - Buddha**

Note that our sense of aliveness is not one of our commonly held five senses. However, it is one of the Organic Psychology 54 natural attraction senses, #54, the instinctive dance of survival, the love to live.

If you cannot find your sense of aliveness, pinch yourself a few times or hold your breath until it hurts or scares you and then start breathing. Dead folks cannot do any of these things, including reading this.

Expand your knowledge about the life of Earth and you by considering the information at http://www.ecopsych.com/livingplanetearthkey.html

> **"All credibility, all good conscience, all evidence of truth come only from the senses."**
> **- Friedrich Nietzsche**

A potential final exam question: *What part of Planet Earth can be proved to be dead? (who, where, how, when)*

A Self-Evident Fact. From its birth on, the Universe and everything in it was and remains alive.

A Supportive Observation: *Show me, people, saying the Universe is dead and I'll show you people missing out on life.*

EXPLORE NOW: VALIDATE. *CONNECT THE CONCEPT, ABOVE, WITH A CONSENTING NATURAL AREA.*
• **SEVMRATCI** the most interesting felt-sense attractions and senses that you find or that find you[40].(Appendix A: Our Fifty-Four Natural Senses and Sensitivities) • Add this process to an experience you have had in your special area of interest: Art, Creative Writing, Music, Yoga, Parenting, Recovery, Addiction, Renewal, etc.
• **Supportive Reading and Activities:** in a natural area read the next chapter in *Reconnecting with Nature* (or approved course book) including doing its activities and journaling them.
• **Journal Response Form:** For personal growth and reference to describe the value of this section, please complete the Appendix B: Revolutionary Wisdom Response Form: Journal Response Form[41].

#10 An Expedition Challenge

Earth is a global life community whose atmosphere acts like a cell membrane and extends far beyond the substantial portion of the Planet. The atmosphere is as important a part of the Planet as are the continents and ocean. We live under it, inside our Planet, not on it

Get consent from a natural place free from poisonous plants, ticks or other dangers to visit it. They are not attractive to your health. Then get down on all fours in this place.

Imagine that you are a microbial cell crawling around in this planet's sizeable global life community. You are, after all in its biosphere and atmosphere. Therefore you are functioning as a bacterium or cell in the Earth organism.

Close your eyes. For five minutes slowly crawl three steps and then, holding your head very still, open your eyes for three seconds. Close them again and repeat this crawling around, eye-opening procedure.

Then, with your eyes shut, for five minutes or longer, feel with your hands, bare feet, toes, body, and head. Sniff out this giant. Listen, taste, sense it well. Roll on it, rub parts of it against the inside of your arms, feel it with your toes and nose, listen to it with your fingers, suck it into your nostrils, crawl on it, smell the scene.

Then, open your eyelids, drink Nature in with your eyes, seek the new and different, be Nature's friend, climb its trees, squint at it out of focus, find different angles, know the grass as a forest.

Challenge: Will you recognize that you are part of all this intelligent relating and vice versa? Can you enjoy your sensations while knowing that they are this omniscient giant, the life of Planet Earth, being you?

Additional Information: http://www.ecopsych.com/54naturepath.html

"One touch of Nature makes the whole world kin." - William Shakespeare

A Potential final exam question: *Where does Nature end and you begin?*

EXPLORE NOW: VALIDATE. *CONNECT THE CONCEPT, ABOVE, WITH A CONSENTING NATURAL AREA.*
• **SEVMRATCI** the most interesting felt-sense attractions and senses that you find or that find you[40].(Appendix A: Our Fifty-Four Natural Senses and Sensitivities) • Add this process to an experience you have had in your special area of interest: Art, Creative Writing, Music, Yoga, Parenting, Recovery, Addiction, Renewal, etc.
• **Supportive Reading and Activities:** in a natural area read the next chapter in *Reconnecting with Nature* (or approved course book) including doing its activities and journaling them.
• **Journal Response Form:** For personal growth and reference to describe the value of this section, please complete the Appendix B: Revolutionary Wisdom Response Form: Journal Response Form[41].

#11 An Expedition Challenge

Read the following daydream and write how it makes you feel:
I purposely imagined that I was the natural world. I was attracted to be a wilderness; I loved that I was perfect. It felt good to be intelligent and alive. I grew bigger and bigger until I became the whole global life community. I was Planet Earth, and I was still alive[50]. I felt just like myself except I could not talk or think as usual. However, I could sense and feel. I felt I was the Planet and I loved being it. I felt that if I got too hot, I'd cool myself by enlarging my cloud cover to reflect heat; I'd get rid of heat-blanketing carbon dioxide by burying it; I'd expand my oceans, or create storms or move my glaciers south.

If the air felt stuffy, I'd swallow carbon dioxide and exhale fresh oxygen. If my oceans got too salty, I'd crystallize out the salt by moving my continents to form warm shallow evaporating seas. If I were hungry or thirsty, I'd eat the sunshine or think up a storm and drink. It was weird, but it felt strong and right. I felt content; I enjoyed living; I felt peaceful.

Challenge: Do you think contemporary science should respect the senses and feelings of this dream but instead usually rejects them?

Do you think Kahlil Gibran's statement, "Beauty is eternity gazing at itself in a mirror" is valid?

Additional Information: http://www.ecopsych.com/54aliveearth.html

A potential final exam question: *Why the scientific name given to Nature's ability to balance itself? (who, where, how, when)*

EXPLORE NOW: VALIDATE. *CONNECT THE CONCEPT, ABOVE, WITH A CONSENTING NATURAL AREA.*
• **SEVMRATCI** the most interesting felt-sense attractions and senses that you find or that find you[40].(Appendix A: Our Fifty-Four Natural Senses and Sensitivities) • Add this process to an experience you have had in your special area of interest: Art, Creative Writing, Music, Yoga, Parenting, Recovery, Addiction, Renewal, etc.
• **Supportive Reading and Activities:** in a natural area read the next chapter in *Reconnecting with Nature* (or approved course book) including doing its activities and journaling them.
• **Journal Response Form:** For personal growth and reference to describe the value of this section, please complete the Appendix B: Revolutionary Wisdom Response Form: Journal Response Form[41].

[50] http://www.ecopsych.com/aliveearth.html

THE WARRANTIED FACT CHECK: Visit the warrantied fact list[51] and insert dates on new facts that you learned or know to this point in the book. Place a check mark on facts you previously dated that you feel have been reinforced or extended.

CHAPTER FIVE: The Web-of-Life Model

THE LIFE OF OUR 54-SENSES

The nature-connecting expedition that this book takes you on recognizes that to be part of a system anything, including yourself, must be in contact with the system otherwise it trespasses or separates from it from lack of communication and troubles is generated. **This is key to explaining and solving Industrial Society's trespasses, disorders, and discontents.**

Without us necessarily being aware of it, the fact is that we have 54 natural senses that continually connect us with the life of Nature/Earth and its well-being, globally and locally[52]. If the stories that guide how we think, sense feel and act omit this connection fact, they steer us to produce the miseries we presently suffer.

Jeff is injured because while he is painting the ceiling, Bill yells, "Hold onto the brush tight," and then borrows the ladder.

We had 54-natural attraction, valid but ignored, natural senses that love to bind us and the brush to the natural world safely.

To become more familiar with this 54-sense, self-evident truth (Appendix A: Our Fifty-Four Natural Senses and Sensitivities). Identify the senses listed there that you have experienced and or seen others experience. Did you ever learn that they continue, for your well-being, "to connect your paint brush to the ceiling"?

Five of these 54-senses, as identified by Aristotle (350 B.C.), are touch, taste, smell, sight, and sound.

---

51 http://www.ecopsych.com/54warrantfact.docx

52 http://www.ecopsych.com/54insight53senses.html

The 49 others include senses of thirst, gravity, hunger, community, excretion, reason, consciousness, love, respiration, literacy, trust, aliveness and humor. Moment by moment the life of each sense loves and nurtures some natural part of our life and all of life.

Note that our runaway society omits to teach us the 54-sense truth about ourselves. However, we experience it, so we know it is a self-evident fact of life. **Our life loves to sense and feel that it is alive.** We call this sensation "survival" or "self-preservation" (senses #52, #54).

Because it is inconvenient to the profits generated by materialism and social power, our leaders do not require that we validate our authentic, 54-sense life experience as scientific fact and modify our laws and education accordingly[6].

When we are socialized to think using 8 of our 54-senses, we operate with 85 percent of our natural intelligence missing, with the IQ 15, that of an idiot. Does that help explain Earth Misery?

> Sherlock Holmes and his sidekick Dr. Watson, an accomplished expert in critical thinking, science, and speculation, went on a camping trip. They pitched their tent in an open area and turned in for a good night's rest. In the middle of the night, Holmes awakened the good Doctor.
>
> "Watson," Holmes queried, "What do you see and what do you deduce?"
>
> Watson rubbed his eyes, looked at the open sky and replied, "I see the sky, and I deduce that among all the planets and stars in the heavens there must be some planets like Earth. Moreover, if there are planets like Earth there is a high probability that there is life out there."
>
> Holmes replied, "WATSON, YOU KNUCKLEHEAD! Somebody stole our tent!"

The Web of Life Model

Because the web of life is held together by natural attractions, in this model each strand of the web is a webstring, and we inherently register 53 of them. They are our 53 natural attraction senses, the 54th being Literacy/stories that we alone enjoy. A unifying activity that portrays the web of life as a spider web is a robust training model[53]. It is described in Part Three of this book and online, too[11].

---

[53] http://www.ecopsych.com/webstrings1000.html

Expand your knowledge about you and your 54-senses by going to an attractive natural area with the list of 54-senses at http://www.ecopsych.com/insight53senses.html. See and label which of the 54 you can find operating in the area as they are working or have worked in yourself. The best way to do this is with others via our Orientation Course ECO 500A http://www.ecopsych.com/orient.html

Additional Information: http://www.ecopsych.com/54ksanity.html

A potential final exam question: *Why is the Webstring Model called a blueprint? (who, where, how, when)*

EXPLORE NOW: VALIDATE. *CONNECT THE CONCEPT, ABOVE, WITH A CONSENTING NATURAL AREA.*
• **SEVMRATCI** the most interesting felt-sense attractions and senses that you find or that find you[40].(Appendix A: Our Fifty-Four Natural Senses and Sensitivities) • Add this process to an experience you have had in your special area of interest: Art, Creative Writing, Music, Yoga, Parenting, Recovery, Addiction, Renewal, etc.
• **Supportive Reading and Activities:** in a natural area read the next chapter in *Reconnecting with Nature* (or approved course book) including doing its activities and journaling them.
• **Journal Response Form:** For personal growth and reference to describe the value of this section, please complete the Appendix B: Revolutionary Wisdom Response Form: Journal Response Form[41].

#12 An Expedition Challenge

In a consenting natural area read aloud the 54-senses in the Affinity-Feeling Network[27] list for ten minutes imagine yourself having no language sense #39 but instead feel and act from the rest of the senses listed. SEVMRATCI that feeling.

Write

1. How your life might feel under these circumstances concerning fun, purpose, community, time, sense of place, stress, and trust.

2. How would you feel about the natural world if you only knew attractive sensations?

Would you be different from the natural world? How?
3. What connection would you feel between your personal desire to be and the natural world's consent for your being?

- While you are thirsty, drink water until it balances your body's need for it. Notice how your thirst feelings naturally disappear. (or do this with your sense of hunger or taste for sweets.)
- The vast sensation-affinity network within and around us helps guide and balance all of life.
- As natural senses, such as thirst, produce contact between entities, they modify and create balance. For example:

  - excretion senses offset intake senses;
  - senses of fulfillment inhibit senses of desire;
  - senses of nurturing offset senses of unfulfilled attraction;
  - senses of place, community and trust modify destructive personal desires.

- The sense of reason helps us sense Nature's balance as a felt logic. For example, it feels sensible to stop drinking when we are no longer thirsty; it feels reasonable to find more attractive relationships when things become too tense or destructive.

  "We have repressed far more than our sexuality: our very organic nature is now unconscious to most of us, most of the time, and we have become shrunken into two-dimensional social or cultural beings, aware of only five of the hundreds of senses that link us to the rich biological nature that underlies and nourishes these more symbolic and recent aspects of ourselves." - Norman Brown

A Potential final exam question: *How many natural senses can you name that you can know and learn from? What purpose or role do they serve?*

The terrifying effects of our abuse, violence, and evisceration of Nature, in and around us, often prevents those of us victimized by it from dealing with the hurtful aftermath that we suffer. Our good experiences in Nature show the real strength we need from Nature is available in the 54-sense fortitude of natural area attractions and that they benefit us as we Grok them. Organic Psychology works because, as Sir Arthur Conan Doyle said: ***"When you have eliminated the impossible, whatever remains, however improbable, must be the truth."***

Each story and rewarding experience in this book helps you reinforce and build wise, abuse-resistant relationships because they are unquestionable:

- In the moment naturally powerful
- 54-senses intelligent
- Self-evident true
- Naturally attractive
- False story resistant
- As alive as Nature

Are you 100 percent Grokked with them? They help us continue to grow as the whole person that we were born rather than remain abused violated or stunted to be a slave to the dollar.

EXPLORE NOW: VALIDATE. *CONNECT THE CONCEPT, ABOVE, WITH A CONSENTING NATURAL AREA.*
• **SEVMRATCI** the most interesting felt-sense attractions and senses that you find or that find you[40].(Appendix A: Our Fifty-Four Natural Senses and Sensitivities) • Add this process to an experience you have had in your special area of interest: Art, Creative Writing, Music, Yoga, Parenting, Recovery, Addiction, Renewal, etc.
• **Supportive Reading and Activities:** in a natural area read the next chapter in *Reconnecting with Nature* (or approved course book) including doing its activities and journaling them.
• **Journal Response Form:** For personal growth and reference to describe the value of this section, please complete the Appendix B: Revolutionary Wisdom Response Form: Journal Response Form[41].

CHAPTER SIX: The Sensory Triad CRL

THE SENSORY TRIAD: CONSCIOUSNESS, REASON, AND LITERACY (CRL)

The accuracy of our story regarding the life we share with Earth is crucial for well-being because it determines the efficacy of our ways and means to reverse the shared misery we produce and impose upon the life of Earth. Using an inaccurate story is like the miserable results of drilling a hole in the bottom of your boat and, seeing the water rush in, you drill another hole in the bottom to let it out.

Beyond reasonable doubt, the information in the previous chapters brings its scientific story to our S-E senses of reason, consciousness, and literacy (Cohen, 2008).

As part of the Big Bang universe, the life of Earth, moment-by-moment, loves to non-verbally produce its own time and space wilderness relationships through at least 53 natural sensitivities that humanity shares. Uniquely, we alone can:

1) register these relationships as nature-disconnected, abstract story narratives, or

2) think and feel with them in either literate-story form and a fundamental, non-story 54-sense relationship with nature.

Both can take place while we are in sensory contact with a natural area, backyard or backcountry. This is true as well when our senses register the sensory life of nature in each other. This phenomenon condenses into the truth that at any given moment while they are in contact with our 51 different senses, our S-E senses of **C**onsciousness(C) sense #43 and **R**easoning(R) sense #42 can resonate, think and act in conjunction with **L**iterate stories(L) sense #39.

Reasonable, evidence-based, 54-sense stories genuinely connect us with Earth/Nature wisdom in a balanced way. They help us produce responsible and supportive personal, social and environmental relationships at any moment.

Nature-disconnecting, belief-based, limited sense stories produce unreasonable relationships. Not being scientifically accurate, they remove us from or dilute the scientific truth of our inherent, Nature/Earth whole-life wisdom that produces each moment's time and space of the Universe. This makes us produce our earth misery. We are like a goat doing brain surgery.

CRL is the acronym/tool that we can use to trigger our S-E sense of

Consciousness(C) sense #43

Reasoning(R) sense #42

Literate stories(L) sense#39

An Organic Psychology process that includes associated self-evident input from our 51 other sense groups is termed ***CRL is CRL-51.***

In our challenge to reverse earth misery and increase well-being, our choice to scientifically apply CRL-51 empowers us to register and be guided by evidence-based, nature-connecting, whole life stories. Our sense of reason recognizes and loves these CRL-51 stories to consciously plug us into Earth's eons of self-correcting, multi-sensory wisdom, in and around us in a natural area. Moment-by-moment, this act in nature, restores balance, purity, and beauty, personally and globally. It strengthens our sustainability, ethics, morality, and wellness. It shapes our posture and attitude into consciously or unconsciously designing our next moment.

CRL-51 enables our distinct sense of reason to register that our outdated, unscientific and often addictive nature-disconnecting stories increase rather than decrease earth misery. Our thinking learns that CRL-51 is the working essence used in the art and science of Organic Psychology for educating, counseling and healing with nature. By applying CRL-51, we organically increase the intelligence of our 5-sense sensibilities, sensitivities, and relationships by 85 percent and we reduce expenditures and conflicts accordingly (Cohen, 2013a). That fundamental story is, "nurture nature."

A potential final exam question: *What makes CRL the most significant contribution of Organic Psychology? (who, where, how, when)*

EXPLORE NOW: VALIDATE. *CONNECT THE CONCEPT, ABOVE, WITH A CONSENTING NATURAL AREA.*
• **SEVMRATCI** the most interesting felt-sense attractions and senses that you find or that find you[40].(Appendix A: Our Fifty-Four Natural Senses and Sensitivities) • Add this process to an experience you have had in your special area of interest: Art, Creative Writing, Music, Yoga, Parenting, Recovery, Addiction, Renewal, etc.
• **Supportive Reading and Activities:** in a natural area read the next chapter in *Reconnecting with Nature* (or approved course book) including doing its activities and journaling them.
• **Journal Response Form:** For personal growth and reference to describe the value of this section, please complete the Appendix B: Revolutionary Wisdom Response Form: Journal Response Form[41].

What CRL story might Nature tell us about how it works? Will it help the balance and beauty in a natural area strengthen us knowing that the love of our personal life is an expression of Earth's appreciation of its life?

What does a natural area convey about this story?

> A plant seed grows into a tree, and upon its "death" it transforms into being other life energies that are attracted to help other living beings similarly survive. That is how the life of Nature/Earth works. Each element individually and collectively is supported; each is a

hologram of Nature's attraction essence. This explains how, over the eons, Nature/Earth's self-correcting optimum of life, diversity, beauty, purity, peace, wellness and cooperation has not produced contemporary humanity's runaway garbage, stress, and abusiveness. Our stories that lead us to abuse the life of Nature/Earth are the culprit we must help CRL address.

The cones of an average Douglas Fir produce at least ten thousand seeds each year, and the tree can live for as many as five hundred years. That's about half a million seeds it produces in its lifetime. However, only one seed's love to grow is needed to reproduce and replace the life of the tree. The remaining seeds are the tree's attraction to feed and support everything else that sustains the tree, including fires and storms. That kind of unconditional love in a natural area is cooperation, not competition. Most of life is born to support the whole of life. The one that is fittest is the most generous and reciprocal, not dominant. We seldom run across that story in our CRL concerning the essence of heart-centered relationships.

Where is your life with regard to the information, above? How does it affect your past and present, your dreams and future?

A potential final exam question: *What CRL story can help us explain how Nature works through survival of the most cooperative? (who, where, how, when)*

EXPLORE NOW: VALIDATE. *CONNECT THE CONCEPT, ABOVE, WITH A CONSENTING NATURAL AREA.*
• **SEVMRATCI** the most interesting felt-sense attractions and senses that you find or that find you[40].(Appendix A: Our Fifty-Four Natural Senses and Sensitivities) • Add this process to an experience you have had in your special area of interest: Art, Creative Writing, Music, Yoga, Parenting, Recovery, Addiction, Renewal, etc.
• **Supportive Reading and Activities:** in a natural area read the next chapter in *Reconnecting with Nature* (or approved course book) including doing its activities and journaling them.
• **Journal Response Form:** For personal growth and reference to describe the value of this section, please complete the Appendix B: Revolutionary Wisdom Response Form: Journal Response Form[41].

CHAPTER SEVEN: Space-Time-Universe

THE LIFE OF THE TIME/SPACE UNIVERSE

> **"All things that come to pass exist simultaneously in the one and entire unity, which we call the Universe. ... We should not say 'I am an Athenian' or 'I am a Roman' but 'I am a Citizen of the Universe.'" -Marcus Aurelius**

ORGANIC TRUTH GLASSES: The Organic Psychology Unified Field GreenWave-54 process.

For reasons you will discover as you continue, the full name given to wearing "organic truth glasses" is the Whole Life Science of Organic Psychology Unified Field GreenWave-54. We will call it "organic truth glasses" in this section of this book.

Once you invent a pair of glasses that let you see better, you can use them to see things you could not see before, including that you can look at the lenses while you wear them. You cannot do this with your human eye because although it can see the world, it cannot see itself. (It can "experience" itself, however, via our 54-senses)

Trust what your eye sees in natural areas as ***being you without your stories.*** It is part of your unique, self-evident truth of those moments that over the eons made your eye.

The evidence-based information gathering process that we use to produce 54-sense connections in a natural area is a "nature-connected truth tool." It acts like a scientific "organic truth glasses technology."

When we wear the glasses, they illuminate the facts that our nature-estranged education, counseling and healing practices usually omit. The glasses enable us to see and make authentic 54-sense contact with the life of Nature's wisdom in natural areas.

As previously noted, while exploring our relationships with nature, researchers wearing organic truth glasses have discovered that that in Industrial Society *we spend, on average, over 95 percent of our time indoors and over 99 percent of thinking, feeling and relating is separated from authentic nature.* This means that our body, mind, and spirit are only connected with Nature **for less than 12 hours/lifetime.**

This enormous planet-familiarity disconnection, abandonment and sensory deprivation from authentic Nature heavily contribute to our Earth misery. It is seldom found in nature-connected societies.

A core scientific fact of our evidence-based, technology-established society is that **since its birth (sic) the life of the Universe has been attracted (loved) to build its own time and space, moment-by-moment, since its Big Bang beginning.** This means that its totality is always a unity reproducing its space and time and growing and available moment-by-moment. This is neither new or startling. Plato observed it circa 460 B.C., and particle physics at CERN validated it in 2012 A.D. See: http://www.ecopsych.com/journalcopernicus.html

> **"The world we have created is a product of our thinking. If we want to change the world, we have to change our thinking."- Albert Einstein**

Today, ***there is still no complete definition of the difference between life and death*** because with the Universe loving to be alive, moment-by-moment, the essence of everything is flourishing in one form or another. Death is just another form of life including the life of Earth and Nature.

> **"Some part of our being knows this is where we came from. We long to return. And we can. Because the cosmos is also within us. We're made of star stuff." - Carl Sagan**

Our nature disconnected reasoning and stories usually reject the idea that the prey in a predator-prey relationship is ever attracted to being preyed upon. However, one part of life is always prey (food) for some other part. The prey is drawn to play a game that strengthens it by removing its weakest parts. This brings its growth into balance so that it does not outgrow its support and be stressed.

Our nature-disconnected reasoning surmises that for the prey to experience its potential "death" as an attraction is unreasonable. We overlook that the traumatic shock of potential death naturally, kindly tranquilizes many of the prey's natural senses, including pain and consciousness, while heightening other transformation senses. Also, the prey is not addicted to a world of new brain stories that its attacker threatens. The prey does not fear the loss of the way it knows life, its Ego or its story world, for the prey has little or none.

At any moment that you know you are alive, Earth and the Universe also must be active for the essences of all things are connected and the same in any time/space moment including what you sense and feel. The life of everything, including you, is attached to what came before it and to what happens after it.

> **"The story of the atoms in our bodies is the story of the universe in a real sense. A proton in the nucleus of an iron atom in a red blood cell may have been born in the Big Bang, passed through several stars, and been flung across the galaxy before ending up at its current place in your anatomy." - Matthew Francis**

We are exposed to, or indoctrinated by, many other creation stories, **but they do not result from the pure science process that we depend on to produce the technologies that support our central way of life today and shape how we think.** Sadly, these technological artifacts are often outdated or inferior, short-term substitutes for the whole and lifelong satisfaction that we would otherwise obtain from the life of Nature.

Our technological additions injuriously disconnect us from the lasting peace of mind people once enjoyed from fulfilling contacts with the deep-rooted life of Earth in the planet and its people as we live its life as humans.

The disconnection we create by our attachments to destructive Earth substitutes abuse or thwart our 54-senses. This produces an uncomfortable, deficit in their energy that can make us feel hurt or abandoned. We excessively want, and we feel that we never have enough. This results in our greed, excessiveness and the misery of our natural resource and happiness deficits. This increases our desire to be part of something bigger, our feeling of loss or lack of wholeness. Our lives sense something loving, supportive and trustable is missing to the point that they emotionally attach to and are manipulated by a simple logo, food or tune. These never fully satisfy for when we feel needy, we do not feel full or adequate happiness.

> **"Technology is destructive only in the hands of people who do not realize that they are the same process as the universe." - Alan W. Watts**

A Self-Evident Fact: Things only exist or happen in the now.
A Supportive Observation: *Show me anything that does not exist in the now, and I'll show you myself denying that you are showing this to me from your experience.*

#13 An Expedition Challenge

Let the life of Nature help you enter the future with the whole life stability and energy of the GW-54 present. Let a natural area help you find your whole-aliveness GW-54 self.

1. Get consent to find an attraction in this area

2. Identify its essence and wholeness by adding "ness" to its name. A rock becomes rockness; a tree becomes treeness, etc.

3. Identify aspects of that rockness or treeness in yourself as you are the GreenWave together. Remember, part of it is that part of you that attracted you select this attraction.

4. Note that you now can locate and rely on the "ness" of natural things in the future.

5. Repeat the above by identifying everything as participating in being, the verb, action or relationship that they are. Accomplish this by adding "ing" to attractions that call you. A rock becomes rocking; a tree becomes treeing

6. Repeat the above by calling what attracts you to be a person; a rock becomes a "rock person" or "one of the rock people."

7. Get to know the life of Earth in and around you as it knows itself.

A. In a consenting natural area, find something that attracts you there and ask it to tell you who and what **it is** without it using labels or names for anything but instead using _______ness, _______ing, or _______person.

B. Repeat the above by asking yourself to tell the natural attraction and yourself who and what **you are** without it using labels or names for anything but instead using _______ness, _______ing, or _______person.

Challenge: Will you rely on the parts of the above that bring you stronger and more centered into the future next moment?

Expand your knowledge about you and the life of our Universe. The stars may be dead, but their atoms are our atoms. Their electron life energy is always alive as them and as us.

Visit a sunlit natural area and do http://www.ecopsych.com/giftearthday1.html there and later with beautiful night sky stars. If we are alive, aren't their atoms flourishing since they originated ours? Do dead things give birth to live things or is it a whole life transition?

A Potential final exam question: *Is it helpful to upgrade the names of the senses into 'ness' or 'ing,' such as thirstness or thirsting? (who, where, how, when)*

EXPLORE NOW: VALIDATE. *CONNECT THE CONCEPT, ABOVE, WITH A CONSENTING NATURAL AREA.*
• **SEVMRATCI** the most interesting felt-sense attractions and senses that you find or that find you[40].(Appendix A: Our Fifty-Four Natural Senses and Sensitivities)

• Add this process to an experience you have had in your special area of interest: Art, Creative Writing, Music, Yoga, Parenting, Recovery, Addiction, Renewal, etc.
• **Supportive Reading and Activities:** in a natural area read the next chapter in *Reconnecting with Nature* (or approved course book) including doing its activities and journaling them.
• **Journal Response Form:** For personal growth and reference to describe the value of this section, please complete the Appendix B: Revolutionary Wisdom Response Form: Journal Response Form[41].

#14 An Expedition Challenge

Gain consent to visit a natural area.

Identify things there that you and your stories know as being dead.

Find attractions and relationship to these "dead things" that show that they are not dead but are just slower or different forms of life as are different species. Google to find examples of something being alive like: are minerals alive? is water alive? is air alive?

Question: If the natural attraction is the **essence of the life** of the Universe, what is death in the life of Earth or the Universe? Can it exist scientifically?

Challenge: Can you apply to yourself and your death what you discover, above?

A Potential final exam question: *Scientifically, where does death exist in nature? (who, where, how, when)*

EXPLORE NOW: VALIDATE. *CONNECT THE CONCEPT, ABOVE, WITH A CONSENTING NATURAL AREA.*
• **SEVMRATCI** the most interesting felt-sense attractions and senses that you find or that find you[40].(Appendix A: Our Fifty-Four Natural Senses and Sensitivities) • Add this process to an experience you have had in your special area of interest: Art, Creative Writing, Music, Yoga, Parenting, Recovery, Addiction, Renewal, etc.

• **Supportive Reading and Activities:** in a natural area read the next chapter in *Reconnecting with Nature* (or approved course book) including doing its activities and journaling them.
• **Journal Response Form:** For personal growth and reference to describe the value of this section, please complete the Appendix B: Revolutionary Wisdom Response Form: Journal Response Form[41].

CHAPTER EIGHT: Unified Attraction Field

THE LIFE OF THE UNIFIED ATTRACTION FIELD

Every aspect of Nature consists of and is held together by the life of the Universe's Big Bang, Higgs Boson, gravity and other attraction fields of the Universe. Albert Einstein identified this life force as the Grand Unified Field of the Universe, and in Organic Psychology, by using 54-senses, the field scientifically updates and includes gravity, electromagnetism, Quantum, String and M Theory of Everything (TOE). The essence they all hold in common is a natural attraction, and it unifies.

Scientists have this past century, validated that Earth and we exist in and from the sequence of the moment-by-moment "now" attraction life of the Unified Field that moment-by-moment continues to birth its time and space as the Universe. If the Universe is alive so is its attraction essence that holds it together and is attracted to grow in diverse ways.

The life of Nature's Universe is not a moment by moment process. Instead, it is a constant flow that self-organizes maintains and corrects its life by lasting attraction balances without using stories. We break it down into being moment by moment to make room for our consciousness to hear, translate it and know it as stories, be they accurate or not, while it continues to flow. This explains why science can not prove anything; it can just show what has not existed up to this moment.

In a natural area, when we register an identified sense whose name we already know, we are best in touch with the life we share with our planet home. For us to be whole, we must register the sense; it's feeling, name, and story together because our personal and collective wholeness includes our unique understanding of Literacy #39. Without this blend, our stories are out of touch with the self-evident facts that our feeling experiences in the natural world provide, and we are addicted to and overwhelmed by our nature-disconnected stories. Therefore CRL is essential in our relationships if we are to stop producing earth misery. It gives Nature a voice.

> **"If you wish to make an apple pie** [or any relationship] **from scratch, you must first invent the universe." - Carl Sagan**

We experience the attraction field on many levels for many things. Doesn't our attraction feel alive? Isn't it alive since we are alive even when we may be attracted to something that is "dead" or we are disturbed enough to want to die?

Attraction unifies the dance of Nature/Earth's eons in every moment. Until you can scientifically prove things are not held together by attraction, it is self-evident that attraction is the essence of everything including you. If you are attracted to live, so is everything attractive to your universal life.

> **"When you have eliminated the impossible, whatever remains, however improbable, must be the truth." - Sir Arthur Conan Doyle**

The only thing in the Universe that may be dead is the story that says something is gone and what that story may have killed in us or destroyed as another species.

In the life of our planet, nothing that is evidence-based has yet been found that is not held together by attraction, and **the aliveness of attraction is the essence of love.** Love is not dead. The love of life, "Biophilia," is alive and attractive.

> **"Love is the only reality, and it is not a mere sentiment. It is the ultimate truth that lies at the heart of creation." - Rabindranath Tagore,** 1913 Nobel Prize (literature)

One can immediately apply Organic Psychology simply by spelling the word love as "luf," "l-u-f" being the <u>L</u>ife of the <u>U</u>nified <u>F</u>ield. Then recognize that it originated in a Big Bang "hugs bosum" rather than the Higgs Boson. :)

We can differentiate between local loves, like a love for chocolate, a tree or a person, and whole-life, Unified Field love (sense #54) by calling the latter, **"love-54,"** the love of loving.

> **"Driven by the (natural attraction) forces of love, the fragments of the world seek each other so that the world may come into being." - Pierre Teilhard de Chardin**

Love-54 acts like a whole-life lens essence of Nature's aliveness. Its truth helps our thoughts, feelings, and relationships be focused on the self-evident clarity of their Unified Field point source, space/time essence of any moment. It acts like a camera lens. The lens then projects that love reality onto our screen of consciousness and memory. There it fills the void and deterioration left by our abusive conquest of the life of Earth. This restoration transforms our discontents and our questionable fixes for them into the happiness of constructive relationships with the life of Earth, around and as us.

Expand your knowledge about you and the life of the Unified Field. Visit an attractive natural area and see if you can find an unknown story that identifies where an attraction there ends and you begin.

Additional Information: http://www.ecopsych.com/54journalgut.html

A Self-Evident Fact[31]**:** Everything is held together by attraction.
A Supportive Observation: *Show me something that is not held together by attraction, and I'll show you someone's story manipulating magnets to make them repulse to each other.*

> **"For more than three decades, the Higgs Boson has been physicists' version of King Arthur's Holy Grail, Ponce de Leon's Fountain of Youth, Captain Ahab's Moby Dick. It has been an obsession, a fixation, an addiction to an idea that almost every expert believed just had to be true." - Stefan Soldner-Rembold**

A NON-SCIENTIFIC STORY: To Be or Not to Be, Self, Meet Yourself.

Can you guess who might have written this page?

> **To start to be, the life of natural attraction was first wordlessly conscious that it was attracted to be, especially when it became aware that it could 'not be' since it already was.**

Because the life of this silent attraction was attractive, its life became a seed that was conscious that it loved to survive by becoming more attractive. It began to be so. It became the 'be'-ginning of beingness.

This was so attractive that it continually strengthened itself by attractively repeating it. It became the process of becoming a more attractive process; practice made perfect.

Attractive parts of the process became conscious that they were attracted to matter, to manifest themselves in the life of energies, particles, matter, materials, time and space.

The attractive parts became conscious of this 'attraction to matter' that they held in common.

They unified and strengthened to orgasm into their Big Bang Universe birth and its unified attraction field(s).

"The pleasure of living and the pleasure of the orgasm are identical." - Wilhelm Reich

The unifying fields enabled them, as a whole, to attach to each other in diverse material, conscious and liveliness. They attractively mattered as Matter with more strength, stability, and attractiveness.

Moment after moment they were attracted into the unified and unifying aliveness of energies, waves, particles, atoms, materials, sensitivities and relationships that continue today in unadulterated 'wild' natural areas of Earth. There everything is a balanced and beautiful, organic 'hologram' of the life of the Universe. Nothing there is repulsive, dead or left out as garbage; things do not get wasted.

When contemporary humanity with its fifty-four senses and its stories recently came into the picture as part of the life of Earth, it invented the fiction of repulsion and negativity. This story helped humanity invade, plant and establish **its narrative that it needed to build artificial tropical environments everywhere it wanted to live.**

I swear by all that is holy that humanity's stories invented or created me as the source and reason for this invasive, story-driven behavior that was heretofore unknown as part of Earth's life. This was, and it remains, not attractive or loving unity. It is abstract, meaning "it pulls things apart." Because it is foreign and run away, you must use my/your sense of reason to stop excessively abstracting and come into balance by reconnecting with my universal love, as of old.

Moment by moment, become more whole (holy) and be more whole-life sensible. In natural areas, learn to 54-sense nurture Nature in and around you and each other. This is what the 'wilderness' of my common sense does, locally and globally, when your nature-abusive stories do not trespass it. It is a sin whenever these stories excessively disconnect you from the natural world, in and around you.

What you continue doing today is eating of the tree of knowledge in my garden and making it into stories, rather than merely enjoying the wisdom of your 54 sensory attractions to the tree and happily celebrating its outcomes. You can effortlessly do the latter since you are the garden. After I first created its land, sea, and web-of-life, it was the garden (us) that created you, humankind, in our image on the seventh day. This sequence matches the discoveries made in scientific particle physics stories.

If I were to visit Planet Earth today, contemporary society would kill me as you did Billy Budd because I cannot help you tell your stories about how I work to your benefit and remind you that you have given me no rights to my life[54]. - **God**

"I believe in God only I spell it Nature." - **Frank Lloyd Wright**

In the science of Organic Psychology, it is S-E, until proven otherwise, that what we call "things" are, in reality, information and natural attraction relationships manifesting themselves as these thing events. This truth applies to sub-atomic relationships, matter, our biological and social humanity including our 54-senses, weather systems, the solar system and beyond (Cohen, 2013). It also applies to any spirituality or religion whose God is unconditional attraction/love/energy to life that is found everywhere.

> **"I believe that the Universe is one being all its parts are different expressions of the same energy, and they are all in communication with each other therefore parts of one organic whole. This whole is in all its parts so beautiful and is felt by me to be so intensely in earnest that I am compelled to love it and to think of it as divine." - Robinson Jeffers**

"My religion's not old fashioned, two plus two make four today as they did in my Lord's time," sang Stewart Hamblen. True. Also true is that in the "Lord's time" Math could only help produce candles, swords, spears and pottery bowls while folks believed the sun and planets revolved around the Earth.

Today, Math helps science produce bulldozers, cell phones, assault rifles, stealth bombers, nuclear missiles, microwave ovens, computers and ballpoint pens **while the mystical and supernatural have yet to reverse our increasing Earth misery.** Organic psychology could reduce it by helping us think and feel from nature Grokking experiences that produced reasonable decisions and activism. We might then use our hands to make organic choices at the ballot box rather than just hold them in prayer.

Where is your life with regard to the information, above? How does it affect your past and present, your dreams and future?

A Potential final exam question: *Why isn't God included in scientific inquiry?* (*who, where, how, when)*

---

[54] http://www.ecopsych.com/einsteingod.html

EXPLORE NOW: VALIDATE. *CONNECT THE CONCEPT, ABOVE, WITH A CONSENTING NATURAL AREA.*
• **SEVMRATCI** the most interesting felt-sense attractions and senses that you find or that find you[40].(Appendix A: Our Fifty-Four Natural Senses and Sensitivities) • Add this process to an experience you have had in your special area of interest: Art, Creative Writing, Music, Yoga, Parenting, Recovery, Addiction, Renewal, etc.
• **Supportive Reading and Activities:** in a natural area read the next chapter in *Reconnecting with Nature* (or approved course book) including doing its activities and journaling them.
• **Journal Response Form:** For personal growth and reference to describe the value of this section, please complete the Appendix B: Revolutionary Wisdom Response Form: Journal Response Form[41].

#15 An Expedition Challenge

Attempt to collect different natural objects in this area and *without actually moving them to* note that attachments bond each entity to being where and what it is. Can you find any entities not subject to some attracting force Love-54? (Note that air remains on Planet Earth because gravity attaches it here. Water stays connected to us when we leave it.)

Love-54 makes the fittest those that supportively transform and cooperate with each other. The fittest provides and has the most support. Can you validate and trust your findings here? You know they are factual for you because you experienced them

"Matter is not a thing at all, it is an event," observed Quantum Physicist David Bohm. Scientists tell us that affinities are universal attractions which make being, i.e., matter possible. The formation of Love-54 attraction relationships helps any entity, including yourself, to be in a more full, stable, and secure way. The higher the number of balanced and supportive relationships an living being has, the more stable is its being, it is a Love-in.

Any specific kind of attraction can be called an affinity. The 54-sense relationships are expressed in you as sensations and feelings. SEVMRATCI the affinity you sense in this area and within yourself

Sit or lie in your attractive natural area for a moment. Relax and let it and your imagination take you to the most significant attractive time and place in your life, the place where you felt most supported and secure with people and your surroundings. Write down each of the feelings and sensations you remember from this space. Are they coming from your imagination or your natural surroundings? Is there a difference in this setting?

> **"Each of us is exploring this thing we call life in our unique way. So is every other form of life, from mountains to every leaf on every tree." —Mellen-Thomas Benedict**

Recognize that this remembered place is attractive to you because many of your life's desires were fulfilled in this time and place.

Challenge: Think of this fulfilling time and place as your psychological home and return here often for supportive Love-54.

> **"Love is the only sane and satisfactory answer to the problem of human existence."**
> **- Erich Fromm**

> **"Love worketh no ill to his neighbor, and therefore love is the fulfilling of the law."**
> **- Romans 13:10 KJV**

A Self-Evident Fact: There is no repulsion or negativity in Nature.
A Supportive Observation: *Show me a person who says Nature contains negatives and I'll show you a person who has nothing to say about it.*

A Potential final exam question: *What part of the Solar System does not consist of love? (who, where, how, when)*

EXPLORE NOW: VALIDATE. *CONNECT THE CONCEPT, ABOVE, WITH A CONSENTING NATURAL AREA.*
• **SEVMRATCI** the most interesting felt-sense attractions and senses that you find or that find you[40].(Appendix A: Our Fifty-Four Natural Senses and Sensitivities) • Add this process to an experience you have had in your special area of interest: Art, Creative Writing, Music, Yoga, Parenting, Recovery, Addiction, Renewal, etc.
• **Supportive Reading and Activities:** in a natural area read the next chapter in *Reconnecting with Nature* (or approved course book) including doing its activities and journaling them.
• **Journal Response Form:** For personal growth and reference to describe the value of this section, please complete the Appendix B: Revolutionary Wisdom Response Form: Journal Response Form[41].

THE ALIVENESS OF MATHEMATICAL TRUTH

Pythagoras (circa 500 B.C.) determined the Universe contains a logical, inherent mathematical order.

Mathematics works because it's pure science helps us accurately abstract and symbolize the moment by moment sequence of the Big Bang attraction process of Nature and Earth. Math consists of the Unified Field attraction sequence story of the numbers
0, 1, 2, 3, 4, 5, 6, 7, 8, 9 because the dance of Nature/Earth's eons is unified by attraction moment after moment.

The sequence of numbers conveys in numerical symbols that everything is attached or attracted to what came before it and to what happens after it. When the core value of any number or relationship is adulterated so is the whole singular truth of mathematics because all numerals are attached to and dependent upon what precedes and follows them.

> "5 is my unlucky number, so it was only worth 4.3," said the Architect in the investigation after the skyscraper collapsed. :)

Similarly, when our stories misrepresent or disconnect us from how the life essence sequence of the unified attraction field has worked since the beginning of time, our thinking and relationships are destructively misguided concerning how the life of our planet works today, in and around us, and unsolvable problems arise.

The inaccuracy of simple mathematics is that it is a story that helps us stabilize and understand a single moment in the life of Nature while in Nature's non-story life every moment is different from the previous moment that birthed it.

The sequence 0, 1, 2, 3, 4, 5, 6, 7, 8, 9, 10 being the core of arithmetic arises from the fact that since its beginning humanity has consistently had ten fingers and that fact attracted us to us to depend upon this phenomenon as a consistent natural truth. To this must be added that pre-human Nature came up with animal life having five fingers and this made sense to increase our survival chances. Note that most vertebrate animals have the remnants of five "finger paws," even 300 million years ago.

The singularity power of 0-1 representing the **whole truth of the Universe in any moment** is demonstrated by it meaning "no or yes" as in "no universe exists" or "our Big Bang universe exists." This "in the moment mathematics" also applies to "right or wrong," "black or white," "now or then," "true or false," and "Democrat or Republican." This is the "yes" or "no" basis of how any intelligence works, including that of a slime mold[18], a computer or our reasoning. It indicates whether something or some story is a reality at the moment (except for "alive or dead" since the moment is alive).

> **Nature**[21], the International Journal of Science notes that compared with most organisms, slime molds have been on the planet for a very long time—they first evolved at least 600 million years ago and perhaps as long as one billion years ago. At the time, no organisms had yet evolved brains or even simple nervous systems. Slime molds do not blindly ooze from one place to another—they carefully explore their environments, seeking the most efficient routes between resources; they can solve mazes. They do not accept whatever circumstances they find themselves in, but rather choose (are attracted to) conditions most amenable to their survival. They remember, anticipate and decide. By doing so much with so little, slime molds[18] represent a successful and admirable alternative to convoluted brain-based intelligence.

Moment by moment slime mold intelligence consists of Unified Field attraction producing either (nothing) **0** or (one) **1** new relationship or attraction. It is the moment that attraction is conscious of what it is attracted to and begins to connect with it and the next moment continues to do so. It's like a spider can sense the vibrations of the web strands that lead to the insect food. This is the same love-54 process that was attracted into the moment it Big Banged itself into being the Universe. It is the vital, attractive relationship-building essence found in sub-atomics and Earth, including us, as well as galaxies and beyond. It is how our intelligence and computer's artificial intelligence work. We are it experiencing its unadulterated self every time we Grok an attraction in a natural area. Grokking organically attaches or attracts to what came before as well as what is now and what will happen after. It makes us congruent with how a caterpillar becomes a butterfly, see **Appendix E**: Tree-ness.

The things in the life of Planet Earth that do not use this intelligence are moments that attach us to stories and technologies that deny or conquer this core wisdom, or artificially replace it and have adverse side effects. We suffer because far too many parts of contemporary society do this.

The Golden Ratio Dance

> **"A fractal is a way of seeing infinity." - Benoit Mandelbrot**

A unique fact of Organic Psychology is that a logical sequence of numbers conveys in symbols that everything is attached or attracted to what came before it as well as connected/attracted what happens after it. This truth expressed mathematically conveys this core of the life of Nature/Earth/Humanity and its dance:

At any given moment anything or event seamlessly consists of three phenomena operating simultaneously. They are (1) what preceded the thing (2) while it is adding what it is to (3) what it will be. This is a natural attraction in the past, present, and future happening all at once. It is the moment that attraction is conscious of what it is attracted to and begins to connect with it.

Numerically, m-m, it looks like this: 0, 1, 1, 2, 3, 5, 8, 13, 21, 34, 55 ad infinitum. This has long been known, but not entirely explained, like the Fibonacci sequence (fractal pattern).

The Organic Psychology explanation demystifies the Fibonacci Golden Ratio [55] (PHI 1.618) of nature. It discloses why the ratio shows up everywhere as an **attractive** spiral or other design, in and around us, from seashells to our finger bone sizes and fingerprints, from hurricanes and spiral galaxies, and DNA nucleotides. We see it as a mystery in our story way of knowing because we don't learn to include m-m 54-sense m-m natural attraction/love in how we learn to think and relate. Each love-54 moment builds on and adds to the previous love-54 moment.

Expand your knowledge about yourself and the life of our Universe. Visit a natural area and see if you can sense and identify examples where Nature is different, moment after moment, or the Golden Ratio can be seen. Is a tree ever the same while the wind is blowing its leaves or a bird lands on it? Are you the same as before you read this or before you saw the tree? A lichen modified a rock during the time you read this, a slime mold[18] solved a problem, and a twig added a leaf.

Every moment of anything is a hologram of its essence-seed becoming more whole-life unique in each moment; this is The Tree of Life in action.

Additional Information: http://www.ecopsych.com/54ksanity2.html

> **"All our knowledge begins with the senses, proceeds then to the understanding, and ends with reason. There is nothing higher than reason." -Immanuel Kant**

A Potential final exam question: *Why is mathematics considered to be pure science?*

---

[55] http://jwilson.coe.uga.edu/emat6680/parveen/fib_nature.htm

EXPLORE NOW: VALIDATE. *CONNECT THE CONCEPT, ABOVE, WITH A CONSENTING NATURAL AREA.*
• **SEVMRATCI** the most interesting felt-sense attractions and senses that you find or that find you[40].(Appendix A: Our Fifty-Four Natural Senses and Sensitivities) • Add this process to an experience you have had in your special area of interest: Art, Creative Writing, Music, Yoga, Parenting, Recovery, Addiction, Renewal, etc.
• **Supportive Reading and Activities:** in a natural area read the next chapter in *Reconnecting with Nature* (or approved course book) including doing its activities and journaling them.
• **Journal Response Form:** For personal growth and reference to describe the value of this section, please complete the Appendix B: Revolutionary Wisdom Response Form: Journal Response Form[41].

CHAPTER NINE: SUNNEH-54 and NNIAAL-54

GREENWAVE-54: OUR ORGANIC TRUTH GLASSES

In the 1950's a Gestalt Psychology counselor at Columbia University produced a diagram that portrayed how we only live, think and act in the present moment and how this is like surfing the crest of a wave.

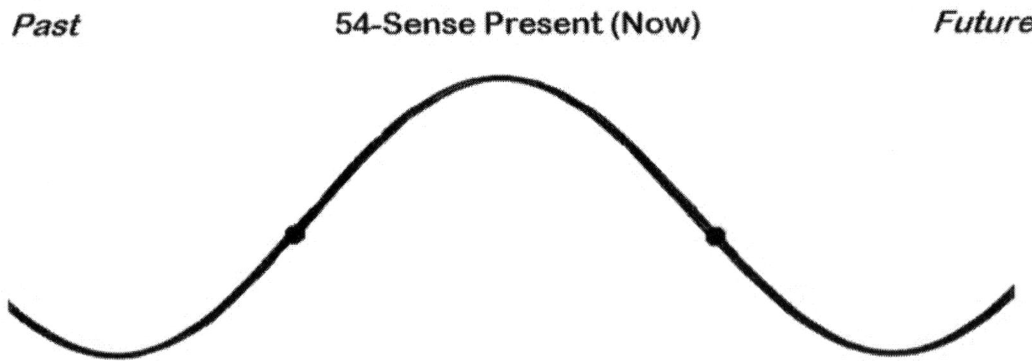

This wave diagram helps us visualize that only in the instantaneous crest peak m/m **Present** can we experience, learn and act in this now moment of our aliveness.

> **"Time past and time future, what might have been and what has been, point to one end, which is always present." -T.S. Eliot**

We can only register and deal with the past and future as stories or memory sensations while we are in or on the "present crest." This is the only time that everything is available to our life including our thoughts, senses, feelings as well as stories, spirit and ideas from the past or future. This is also the moment of the universe, Love-54 Unified Field making its own time and space.

The facts, above, about the crest, hold true if folks do or don't believe in or experience God. They can only do either in and on the immediate wave crest "now."

The *now* is the immediate moment that you can change something, or make choices, or act.

Whatever 54-sense registers in the now moment on the crest is the most significant truth that you can trust or experience at that time. Everything else is fiction or non-fiction information that may come into consciousness as a story that occurs in the now crest moment.

> **"Without the concept of time, there is only the wholeness of nowness." -Stanley E. Sobottka**

Greenwave-54 acts as a science and technology power tool that works similarly to how the automobile has replaced walking. It is like a magnifying glass that takes the energy of the sun and focuses it into one energy spot whose intense heat has the power to start a fire. Similarly, looking through the magnifying glass enlarges part of the view considerably. This brings a small area or moment into sharp focus where we can relate to the essence of its ways. It unifies by bringing the whole as well as its core into the moment.

The value of GreenWave-54 is that it empowers us to find the higher wholeness of our life as we register its essence through our 54-senses on the crest including consciousness, reason and literate/stories (CRL). They, too, as part of anytime/space Unified Field moment, are present in the now. This helps us see the world through organic truth glasses that are often missing in Industrial Society's nature-disconnected relationships, education, counseling, and healing.

Grokking 54-sense contact with the crest of the wave along with CRL brings into view our common culture's critically missing element, the

> Now, Nameless, Intelligence, Aliveness, Attraction Love-54 **(NNIAAL-54)** attributes of Science, Universe, Nature, Earth, and Humanity **(SUNEH-54)**.

As a fast thinking and feeling remedy for many miseries, it can be helpful for our sense of reason #42 to bring all these elements into our awareness quickly by remembering the CRL acronym **SUNEH/NNIAAL-54.**

Some folks add NNIAAL(N) to SUNEH-54 to make it **SUNNEH-54.**

With practice, reasonably Grokking a natural area attraction while thinking the acronym NNIAAL or SUNNEH-54 can bring the researched truth of the wave crest into play and reinforce Love-54. Mathematically this becomes our oneness with Nature at that moment, an arithmetic of love, a technology of behavior. Applying it helps the whole of life, in and around us, self-correct the destructive half-truths and discontents that Industrial Society teaches us to produce but not how to cure.

Additional Information: http://www.ecopsych.com/54earthstories101.html

A Potential final exam question: *Is riding the GreenWave crest a reasonable substitute for God if we expect to curtail our destructive ways? (who, where, how, when, why)*

EXPLORE NOW: VALIDATE. *CONNECT THE CONCEPT, ABOVE, WITH A CONSENTING NATURAL AREA.*
• **SEVMRATCI** the most interesting felt-sense attractions and senses that you find or that find you[40].(Appendix A: Our Fifty-Four Natural Senses and Sensitivities) • Add this process to an experience you have had in your special area of interest: Art, Creative Writing, Music, Yoga, Parenting, Recovery, Addiction, Renewal, etc.
• **Supportive Reading and Activities:** in a natural area read the next chapter in *Reconnecting with Nature* (or approved course book) including doing its activities and journaling them.
• **Journal Response Form:** For personal growth and reference to describe the value of this section, please complete the Appendix B: Revolutionary Wisdom Response Form: Journal Response Form[41].

MAKE YOUR GREENWAVE-54 SUNLIGHT ZIPPER

Get a sheet of the blue paper and a sheet of the orange paper and paste or tape them together to become one single true-blue/orange page. This is like a zipper combining two different parts of a coat to produce the unified value of a full coat.

The blue is called **true-blue** because it and everything placed on it **is evidence-based and scientifically substantiated.**

You have created for yourself a new technology, a whole life tool, that the world may have never seen before. Its authority and authenticity can be used as a talking stick, a magic wand, a mallet, scepter, baton or nature textbook to produce life-sustaining balance and integrity.

Imagine that the orange side of the page is the now moment of the Universe on which appears all the **nature-substitute or disconnecting stories** of and about the world that have been written or spoken since the beginning of humanity.

Imagine that on the true-blue side of the page are all the scientific stories that **connect** contemporary thinking and people with the life of nature, with how and why the beauty of the natural world's self-correcting perfection works in and around us. Recognize that society's assault on this true-blue side produces the earth misery we suffer.

On this true-blue side appears the names of the 54-senses as well as the GW-54 processes in this book that work for you now, and later as you proceed.

#16 An Expedition Challenge

The true-blue facts include: (*which of them, if any, seem unfamiliar to you?*)

- We have fifty-four natural senses that consciously bind us in unity with the natural world.
- NNIAAL and SUNNEH-54 and **CRL** are singular moments in time.
- Webstrings and NSTP are GreenWave-54 in action.
- Self-evidence is undeniable, experienced, sensory fact that omits the mystical and supernatural.
- We have "two bodies," verbal/social and nonverbal/Earth.
- We need an appropriate "connector" to join our two bodies in the balance and beauty of the whole life of our planet.
- The sense and sensation of thirst is a scientific fact.
- We have real-time 54 natural senses; they are not hallucinations
- We live in as part of, not on, the life of Planet Earth.
- The Unified Field is attracted to produce the "now" time/space singularity of the Universe, moment-by-moment.
- SEVMRATCI helps to strengthen our 54-sense sensibilities.

- 2 + 3 = 5 mathematics is true as abstract absolute fact.
- The unified attraction Love-54 field exists everywhere.
- Attraction is conscious of what it is attracted to so it knows what it will attractively attach to.
- Grokking experiences are m/m self-evident, scientific facts.
- Life and death are different forms of natural attraction aliveness; all things are alive
- Valid science can't omit including the evidence of its adverse effects as facts it must consider.
- The life of Planet Earth/Nature is also our life. What happens to it, happens to us
- The purpose of life is to support life and its survival.
- We can't learn how to reconnect with nature from folks who do not know how to do this.
- We need to scientifically solve the problems that we scientifically produce because scientific stories are foreign to speechless Nature.
- The sequence of the seamless, Unified Field continuum makes anything attached to all that came before it and to all that comes after it.
- An orgasm is the Big Bang repeating its pure birth as the moment of new life for the parent and child.
- The **CRL** Conscious, Reason Literacy sense *triad* (#43, #42 #39) addresses the point source of unsolvable problems.
- The non-literate, speechless, aliveness of Earth, the fountainhead of authority in its life, has not written books and articulated stories about how its homeostatic perfection works. We need to add GW-54 *authentic natural area contact experience* to our personal or collective lives to gain accurate and practical story information for the problem-solving.

Imagine that the glue or tape that you used to hold the pages together is the unifying attraction energy field of the Universe, m/m Love-54 **as well as it is you attracted to applying the Love-54.**

You do not have to imagine this. It is as accurate for this glue as it for anything else that is held or holding together including ourselves putting together and using this true-blue/orange paper as a CRL-51 tool.

As a singularity, everything in the Unified Field is its now time and space to which CRL-51 connects you.

> **“I know that what you call "God" actually exists, but differently from what you think: as the primal cosmic energy in the universe, as your love in your body, like your honesty and your feeling of nature in you and around you.” -Wilhelm Reich**

Recognize that it is your now attraction to the words on this page that attracted you to glue the true-blue and orange pages together. It is also you being both true-blue and orange ways *of* relating, the orange being with nature-disconnecting words or stories.

- Recognize that my attraction to place these words on this page here made them available to satisfy your desirability to read them and to glue the true-blue and orange pages together.
- Recognize that reasonable scientific attraction "raw materials" elsewhere made this possible this page as a technology. It has also established its technicians and humanity etc. and ad infinitum since the beginning of time.
- Scientifically, this all has taken place because **an essence of the GreenWave-54 is that "Attraction/Love-54 attach everything to all that came before it and to all that follows it."**

The natural world is how attraction civilized itself to be what we call Nature, and it includes us.

Stories that destructively disconnect us from the natural world need to be corrected by nature's civilization process. The GreenWave CRL-51 makes this possible.

This one thin true-blue and orange page you have made is the last page in "the book of time."." That book simulates a "SERI" or "ALESIS" or "ECHO" question- answering product that is knowledgeable about time in Nature and its material presence and space at this moment.

This book is the world manifesting itself as a thin, space/time, true-blue and orange piece of CRL-51 paper, moment-by-moment, page after page. It is SUNNEH-54 in action producing itself, the paper and us. It is like the last page in an expedition book that is adding another page this instant and every new moment.

Part of a true-blue story about the materials that the glued-together sheets of paper consist of says that the paper was born as a tree from a seed nourished by the biosphere and SUNNEH-54.

Hundreds of other stories have and could be told about the paper as well and a lifetime spent sorting them out. However, being stories, whatever that entailed would mostly be written in the orange, nature-disconnected side of the CRL-51 paper because nature does not tell or understand stories. **It's immediate; true-blue 54 sensibility attractions manifest themselves as things like true-blue trees, our 54-senses and us** along with everything else.

Where is your life with regard to the information, above? How does it affect your past and present, your dreams and future?

A Potential final exam question: *Can CRL on the GreenWave crest connect us to the core attraction being conscious of what it is attracted to in any given moment? (who, where, how, when)*

> **A Self-Evident Fact.** GreenWave-54 makes an essential contribution to increasing personal, social and environmental well-being. (See Appendix A: Our Fifty-Four Natural Senses and Sensitivities)
> **A Supportive Observation:** *Show me those who say GreenWave-54 is not significant and I'll show you why the Little Blue Macaw and West African Black Rhinoceros recently went extinct.*

EXPLORE NOW: VALIDATE. *CONNECT THE CONCEPT, ABOVE, WITH A CONSENTING NATURAL AREA.*
• **SEVMRATCI** the most interesting felt-sense attractions and senses that you find or that find you[40].(Appendix A: Our Fifty-Four Natural Senses and Sensitivities) • Add this process to an experience you have had in your special area of interest: Art, Creative Writing, Music, Yoga, Parenting, Recovery, Addiction, Renewal, etc.
• **Supportive Reading and Activities:** in a natural area read the next chapter in *Reconnecting with Nature* (or approved course book) including doing its activities and journaling them.
• **Journal Response Form:** For personal growth and reference to describe the value of this section, please complete the Appendix B: Revolutionary Wisdom Response Form: Journal Response Form[41].

WHAT TO DO

The Single Page Technology

In a people-built area hold that thin sheet of CRL-51 orange/true-blue paper in front of you sideways so you can see how thin it is. Then imagine all the built environment and in words and artifacts that you see and feel around you now are sucked into and become stories on the orange side of the paper at this moment. Note that they are continually adding themselves to it, as are your stories. The orange side is our sense of literacy/stories, sense #39, at this moment.

We spend, on average, **all but twelve hours of our life in Orange** consciously thinking and feeling in our nature-estranged human-built, artificial orange side and adding our **nature-disconnected** *experiences to it.*

In a natural area hold that that thin sheet of CRL-51 orange/true-blue paper in front of you sideways so you can see how thin it is. The **true-blue** side of the paper represents the 54 love-sense scientific histories of the natural world since the beginning of humanity **continuing to be its natural self at this moment.**

To our loss, on average we spend **less than one day of our life in true-blue tune** with this side. The nature-connected stories in this experience accurately say that they represent "contactless," "relation shipping" and "lifeing" in the natural area.

The Yellow CRL Sun Zipper

> **"Making the simple complicated is commonplace; making the complicated simple, awesomely simple, that is creativity." - Charles Mingus**

#17 A Yellow Zipper Expedition Challenge

When our CRL-51 story makes the time and space to hold the thin, immediate moment, true-blue/orange paper in front of us sideways, it is like our nature-disconnected stories, and our natural world selves are the two separated sides of this unified moment in the sun.

Now, using the sunlight's illuminating energy, **we construct a yellow zipper technology, one that zips everything in Nature together in the light of day so we can see it.**

> Yellow mixed with True Blue becomes Green. The Green of the field, forest, and algae.
>
> True Blue mixed with Orange becomes Brown. The Brown of soil, roots, bark, and excrement-food.
>
> We use nature's yellow sun zipper to bring the orange and true-blue sides together into the wholeness of CRL. These ***unified stories appear in green or brown as GreenWave-54 on each side of the true-blue/orange page.*** The GreenWave-54 then continues into the next moment in unity.

The fact that our lives **pay little attention** to living and learning to use the sunlight CRL **yellow zipper to connect and beneficially transform our orange and true-blue stories and experiences into GreenWave-54** explains how and why we produce and suffer our disorder miseries.

In technological reality, both sides of a zipper are made up of the same shaped snaps. By scientific human design, the zipper's technique connects the snaps by offsetting one side of the zipper a bit so that it slips into and locks itself between the snaps on the other side rather than gets blocked by them. This unifies the zipper and garment.

Similarly, by riding the GreenWave CRL-51 **we yellow zipper shift our thoughts and relationships from being orange nature disconnected isolation and its miseries to GreenWave-54 nature-connected green wholeness and the happiness of its well-being, in and around us.**

Our expedition challenge in a natural area is to shift into this "zipper page" moment by listening to the true-blue side stories that ask us to Grok this natural area and write our personal experiences from doing this here in GreenWave zipper connectedness. The Grokking produces areas and statements that **we identify in our journal as being green.**

Also, we can take nature-supportive true-blue stories from our orange journal statements and make them green when we know **they are accurate because we have experienced them in the authenticity of the natural area and we choose to benefit from their accuracy.**

Once we do the above, we promise the area and ourselves to start all our stories and relationships with our truth, our yellow zipped GreenWave stories. We do this to help stop the assault on nature by orange stories. This brings our lives and the world into balance m-m.

We can repeat this process and strengthen these GreenWave love-54 stories by repeatedly connecting this thin orange/true-blue page and our lives **with the life of any natural area at any given moment.** For example, in the sky, true-blue, unpolluted "airness," "clouding" and "bird "people." are attracted and delighted to help us with doing this because they know our GreenWave stories will support their lives, too.

Additional Information: http://www.ecopsych.com/54counseling.html

A Potential final exam question: *What sense makes sense of the other 53-senses by validating yellow zipper experiences? (who, where, how, when)*

EXPLORE NOW: VALIDATE. *CONNECT THE CONCEPT, ABOVE, WITH A CONSENTING NATURAL AREA.*
• **SEVMRATCI** the most interesting felt-sense attractions and senses that you find or that find you[40].(Appendix A: Our Fifty-Four Natural Senses and Sensitivities) • Add this process to an experience you have had in your special area of interest: Art, Creative Writing, Music, Yoga, Parenting, Recovery, Addiction, Renewal, etc.
• **Supportive Reading and Activities:** in a natural area read the next chapter in *Reconnecting with Nature* (or approved course book) including doing its activities and journaling them.
• **Journal Response Form:** For personal growth and reference to describe the value of this section, please complete the Appendix B: Revolutionary Wisdom Response Form: Journal Response Form[41].

Validation

Revolutionary Wisdom uses the same practical logic and procedure that is used in updated research to solve unsolvable questions or problems from the past. For example, today, investigators rely on the recent discovery that DNA can help us more accurately identify individuals, other forms of life and past relationships. By reasonably applying today's DNA identification procedures to historical events, researchers discover new truths about the past. The same reasoning is used in the sciences of Geology and Archeology, we apply today's advanced information and logic to previous events.

Similarly, this book adds a new tool, a GreenWave-54 (CRL-51) expedition to increase our understanding of nature's ways and how to supportively relate to them.

#18 An Expedition Challenge

Take the **CRL Zipper** orange/true-blue paper activity to a natural area, and in that singular moment of the Unified Field apply it to the most attractive things that you have a journal written and read so far in this book. Do this also with what you found most challenging. Discover if and how this helps you and others validate the book along with your experiences and responses.

GW-54 helps us examine and interpret sensory evidence from the past and present, evidence that has seldom been recognized or appreciated because the ways and means of GW-54 and its 165 nature-connecting activities have not been available to help us in this quest.

GW-54 enables us to recognize how the results of using GW-54 contribute to the well-being of all as it validates that GW-54 has merit and is available.

GW-54 is trustable because it is built on direct sensory evidence from contact with the Unified Field of nature, the real thing, in people, places, and things. Reminder: a central concept in scientific methodology is that all facts must be empirical, or empirically based, that is, *dependent on evidence or consequences that are observable by the senses.*

A Potential final exam question: *What special attachment value do scientific methodology, the Big Bang, mathematical logic and the Unified Field have in common?*

EXPLORE NOW: VALIDATE. *CONNECT THE CONCEPT, ABOVE, WITH A CONSENTING NATURAL AREA.*
• **SEVMRATCI** the most interesting felt-sense attractions and senses that you find or that find you[40].(Appendix A: Our Fifty-Four Natural Senses and Sensitivities) • Add this process to an experience you have had in your special area of interest: Art, Creative Writing, Music, Yoga, Parenting, Recovery, Addiction, Renewal, etc.
• **Supportive Reading and Activities:** in a natural area read the next chapter in *Reconnecting with Nature* (or approved course book) including doing its activities and journaling them.
• **Journal Response Form:** For personal growth and reference to describe the value of this section, please complete the Appendix B: Revolutionary Wisdom Response Form: Journal Response Form[41].

THE GREENWAVE CONTRIBUTION

As stated in the introduction, our leaders expertly identify the problems that face us, but we increasingly suffer because they do not give us the tools that we need to correct them. Organic Psychology GW-54 is an already noted form of Applied Ecopsychology and Conservation Psychology. Is the field of Ecopsychology or any other field going to accomplish what needs to be done if GW-54 is not applied to the things Ecopsychology already does?

By *omitting GW-54,* for any person who lives in our planet's deteriorating life, it appears as if:

- Self-evidence is not self-evident true-blue.
- 2 + 3 = 5 is not consistently true as an abstract fact.

- The sense of thirst is an illusion, not a scientific fact.
- 46 of our senses are hallucinations.
- The Unified Field does not exist.
- Grokking is not scientific.
- Science can omit including the evidence of its adverse effects as facts and still be valid.
- We do not yet know if the life of Planet Earth exists.
- We can learn how to reconnect with nature and come into balance with folks who do not know how to do this.
- We do not need to scientifically solve the problems that we scientifically produce.
- Love-54 does not attach everything to what came before it and to what comes after it.
- Orgasm is not a Big Bang, a simple restart of sensory life.
- Miraculously the non-literate, speechless, aliveness of Earth, the fountainhead of authority in its life, has written books and articulated stories about how its homeostatic perfection works, so we do not need to add GW-54 **authentic contact experience** with Nature to our personal or collective lives to obtain accurate relationship information.

GW-54 is a revolutionary tool in that it lets Nature demonstrate to any individual where its absence is producing an abuse or problem in their life, and its presence can help them solve that situation. This process creates the well-being of GreenWave relationships on personal and community levels, locally and globally.

- GreenWave folks share, support and strengthen each other as nobody else can because while that know-how is omitted in the contemporary world, GW-54 people are motivated and can use, share and live it as well as teach it to others.
- The missing GW-54 phenomenon in our central way of thinking and relating is an expedition into restoring, in and around us, what has been unnecessarily injured or removed by Industrial Society.
- In GW-54 each person or group carefully builds and expands their own reliable and safe GreenWave playground. It is their sensory, nature-connected sphere of influence with other people. Doing this enables them to reasonably build supportive relationships by encouraging their love-54 playground to blend in attractive ways with the playgrounds of others consensually.
- GW-54 unity is accomplished via real-time GW-54 meetings in conjunction with online group expedition education.

GW-54 is the missing organic tool that we need to remedy our personal and global Earth misery. **Grokking its CRL-51 crest in any situation reasonably helps us bring our destructive attachments into whole life, sustainable balance.**

As you continue learning and applying revolutionary wisdom, it strengthens your ability to enjoy GW-54 love-54 benefits alone and with others. You are welcome to change parts of your name to GreenWave if that will help your identity in this endeavor.

> **"For small creatures such as we, the vastness is bearable only through love." -Carl Sagan**

Where is your life with regard to the information, above? How does it affect your past and present, your dreams and future?

EXPLORE NOW: VALIDATE. *CONNECT THE CONCEPT, ABOVE, WITH A CONSENTING NATURAL AREA.*
• **SEVMRATCI** the most interesting felt-sense attractions and senses that you find or that find you[40].(Appendix A: Our Fifty-Four Natural Senses and Sensitivities) • Add this process to an experience you have had in your special area of interest: Art, Creative Writing, Music, Yoga, Parenting, Recovery, Addiction, Renewal, etc.
• **Supportive Reading and Activities:** in a natural area read the next chapter in *Reconnecting with Nature* (or approved course book) including doing its activities and journaling them.
• **Journal Response Form:** For personal growth and reference to describe the value of this section, please complete the Appendix B: Revolutionary Wisdom Response Form: Journal Response Form[41].

Alive in This Moment Reality

The *Revolutionary Wisdom* chapters that follow reinforce and expand the value of Grokking the GreenWave-54 crest and your SUNEH/NNIAAL-54 literacy and Love-54 that supports it. The pages accomplish this by recognizing that we cannot utilize self-evident experience from the future because it has not yet occurred. We have not authentically registered its space/time reality. What we have and are is the Unified Field moment of the crest. Below, diagrammed accordingly, the future is omitted. The image is self-evident accurate rather than future-misleading.

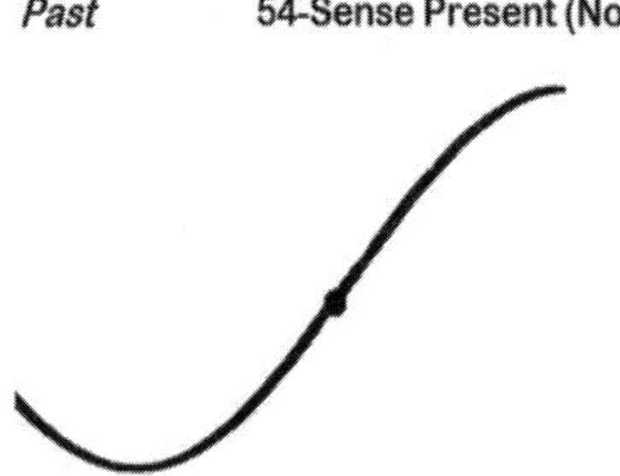

This diagram is a revolutionary update. In it, we can see that as we move into the future on the GreenWave, we can do it with GreenWave-54 ways and stories that we create on the CRL-51 crest accompanied by the GW-54 ability to free ourselves from our attachments to negative and outdated stories, past and present. It is our contemporary knowledge consciously engaged in and benefitting from love-54 slime mold intelligence.

In the original full wave diagram, the future was already bonded as a story in our mentality. This meant that it could exist in the now and easily promote its already indoctrinated ways into the future. This lets our old stories continue to misguide us and produce the earth misery we must now transform into personal and global well-being. Many of these stories become institutionalized[4] and difficult to change when they are not on the GW-54 crest.

A Potential final exam question: *Why is a relationship's background as necessary as the relationship itself?* (what, *who, where, how)*

EXPLORE NOW: VALIDATE. *CONNECT THE CONCEPT, ABOVE, WITH A CONSENTING NATURAL AREA.*
• **SEVMRATCI** the most interesting felt-sense attractions and senses that you find or that find you[40]**.**(Appendix A: Our Fifty-Four Natural Senses and Sensitivities) • Add this process to an experience you have had in your special area of interest: Art, Creative Writing, Music, Yoga, Parenting, Recovery, Addiction, Renewal, etc.
• **Supportive Reading and Activities:** in a natural area read the next chapter in *Reconnecting with Nature* (or approved course book) including doing its activities and journaling them.

- **Journal Response Form:** For personal growth and reference to describe the value of this section, please complete the Appendix B: Revolutionary Wisdom Response Form: Journal Response Form[41].

#19 Expedition Challenge Activity

Increasingly validate your ability to contribute to personal, social and environmental well-being.

You now have the expertise and natural ability to **invoke GreenWave-54.** It's **fifteen-point scientific facts** (see the breakdown, below,) qualify you, at any moment, to trust, register and participate in their authenticity.

The challenge here is to apply the indented A-D statements, below, and get some experience in using GreenWave-54. You accomplish this by **utilizing A-D to your responses** to this book's preceding Introduction, Chapters 1-9, and whatever else in the chapters that you want to examine or challenge. By **invoking GreenWave-54, this way** discovers if or where what you have written, or other areas, **make 54-senses to you as well as begin to heal your abused senses.**

> **A.** Self-evidence from our inherent natural wisdom is irrefutable because it begins directly in our body, mind, and spirit, fifty-four natural attraction senses that our sense of consciousness can accurately register and communicate through our senses of reason and language (CRL).
>
> **B.** In a natural area, the moment by moment attraction energy of the Unified Field being conscious of what it is attracted to, holds the nonverbal, living attraction core of all things together in intelligent, homeostatic equilibrium, from sub-atomics to the life of Earth's Web-of-Life attraction, to circling the life of the sun without producing garbage.
>
> **C.** Reasonably engaging in purifying GreenWave-54 moments and their past and future attachments deactivate the catastrophic effects of extreme disconnectedness and let the wisdom of GreenWave-54 experiences help us produce a safe, 54-sense, life platform that guides us into renewed, whole-life moments of conscious and un-warped relationship evolution.
>
> **D.** GreenWave-54 stabilizing sanity and satisfaction happiness m-m motivates us to thrive by living and teaching GreenWave-54 as a core of our personal experiences,

relationships, and livelihood

The rationale validation and defense of GreenWave-54 is available in the books "Reconnecting with Nature," "With Justice for All" and "Revolutionary Wisdom" by Michael J. Cohen, Ed.D., Ph.D. [And online at
http://www.ecopsych.com/journalaliveness.html
http://www.ecopsych.com/GREENWAVE.docx

If you want to read a ground-breaking Ph.D. Proposal and Dissertation that identifies the benefits of applying A-D to our abuse of Nature, in and around us, please go to http://www.ecopsych.com/journalproposal.html

A breakdown of the A-D fifteen-point scientific facts that you can invoke:

1. Self-evidence from our inherent natural wisdom is irrefutable because it begins directly in our body, mind, and spirit,
2. fifty-four natural attraction senses that
3. our sense of consciousness can accurately register and communicate through our senses of reason and language (CRL).
4. In a natural area, the moment by moment
5. attraction energy of the Unified Field holds the
6. non-verbal, living attraction core of all things together in
7. intelligent, homeostatic equilibrium, from sub-atomics to the life of Earth's Web-of-Life attraction to
8. circling life of the sun without producing garbage.
9. Reasonably engaging in purifying GreenWave-54 moments and their past and future attachments
10. deactivates the catastrophic effects of excessive disconnectedness and lets
11. the wisdom of GreenWave-54 experiences help us produce a safe, 54-sense, life platform that guides us into renewed, whole-life moments of
12. logical and un-warped relationship evolution.
13. This stabilizing sanity and satisfaction happiness motivate us to thrive by living and teaching GreenWave-54 by applying it in our
14. personal experiences, relationships, and livelihood.

15. The rationale validation and defense of GreenWave-54 are available in the books "Reconnecting with Nature," "With Justice for All" and "Revolutionary Wisdom" by Michael J. Cohen, Ed.D., Ph.D.

Resources: http://www.ecopsych.com/journalaliveness.html and http://www.ecopsych.com/GREENWAVEBETA.docx

A Potential final exam question: *What might prevent you from invoking the GreenWave on some level (who, why, when, where, how.)*

Organic Therapy

The phenomenal costs and detrimental impacts of Earth misery[5] suggest that the art and science of its GreenWave-54 Unified Field CRL can be the most respected and requested technology of tomorrow if we expect to have a reasonable future. This is because applying its Organic Psychology process to any endeavor makes the latter significantly more popular, organic, happier, healthier and unified while reducing its costs and destructive effects. Enough people engaged in accomplishing this could shift the life of Earth to produce additional species, habitats, and support for humanity so that the life of Earth can come back into balance.

The key to organic transition is to learn how to therapeutically transform disparaging aspects of our society's stories into the benefits of Organic Psychology relationships. Every one of our nature-disconnected labels and stories can be replaced by GreenWave-54 interactions in ourselves and our sphere of influence in society. This is already occurring where GreenWave connections with nature are the source of higher power[35] in 12 step addiction programs and where they are added to art[56], yoga, parenting, healing, and counseling. It makes any whole enterprise therapeutic rather than strengthen citizens whose relationships increase rather than reverse our runaway earth misery.

Organic Psychology is already proving helpful where trained Organic Psychology facilitators, "Earth Ambassadors" or "Earth Avatars" are presently bringing life back into balance by invoking GreenWave-54:

- Education, therapy and mental health
- Physical health, medicine, life and health insurance
- Parenting, welfare, and social services

[56] http://www.ecoart-therapy.org

- Personal, social and international relationships
- Environmental protection, sustainability, and climate change
- Individual, family and spiritual relationships
- Politics, industry, and employment
- Community development and cooperative relationships
- Nature-deteriorating philosophies and institutions
- Imprisonment and military services
- Weddings, meetings, and funerals

The remaining three parts of this book and course validate, reinforce, accredit and enhance what you now already know so that you can expertly apply it.

At this equinox date, December 21, 2017, the completed chapters for this book will **help your story create a more reasonable and happier personal playground for yourself** by empowering you to use CRL-51 to build it to and help others do the same.

This chapter material is now posted, along with parts of the chapters, above, in our GreenWave-54 Comprehensive Inventory: http://www.ecopsych.com/journalaliveness.html

WARRANTIED FACT CHECK: Visit the warrantied fact list[51] and insert dates on new facts that you learned or know to this point in the book. Place a check mark on facts you previously dated that you feel have been reinforced or extended.

PART FIVE

The History and Application of GreenWave CRL

"What this course has enabled me to do is put the pieces of my disconnected puzzle together.

I knew I was depressed but did not understand why.

I knew I was anxious but could not find a reason to be.

I knew I should be satisfied with my status in life but carried a huge unexplainable empty feeling with me. I had in times of very depressed moods instinctively sought relief in nature, but like someone who had lost the puzzle box cover, the pieces were just too hard to put together.

I feel that I have found the lost puzzle box cover and the big picture can be seen. Pieces are falling into place quite naturally, and a whole New World has been revealed.

Now I not only know how to describe my own natural senses but to listen and learn from those of others.

I no longer "schedule in" time for sensory attractions. I "thirst" for them daily and recognize that I have to quench that thirst in order for my body to be fulfilled.

This course promotes in its members a healthy "fix"; that is, participating in the readings and activities with other interested and involved members gives your mind and body what it needs. You feel so good you naturally crave more of these experiences on your own."

SPECIAL NOTICE

Put PART 1-4 to work. Mentor others through this course to help your and their relationships, expertise, and livelihood enjoy organic wellness.

It makes sense that by bringing our 54 natural senses[27] into play the stories and experiences in this book integrate our life with the life of nature in a natural area. This occurs because in congress our 54 sensibilities detect stimuli on mechanical, thermal and chemical levels that strengthen human and global well-being and survival in balance.

Traditionally, our 54-senses and their beneficial effects are scientifically overlooked yet their restorative powers are irrefutably self-evident and valuable facts of life. Most of us have experienced this to some extent while in a natural area, even just for a short time. Our body mind and spirit register these benefits directly. We temporarily feel relaxed, renewed and belong. We are whom we were naturally born without being socialized by nature-disconnecting stories that warp or injure our senses.

Our lives presently suffer because our 54-senses gift to our life in balance is the crucial omitted element of our doctrines, philosophies, institutions, nationalities, psychologies, economics, relationships, academics, sciences, and spirituality.

By enabling us to experience and validate the how, what, why, when, where and meaning of our 54 natural senses, the process of this book unifies humanity with itself and the natural world on personal and global levels.

In this book's procedures attractive truths reasonably find and bind like magnets naturally attract each other in universal ways. For this reason, we offer subsidized doctoral degrees to individuals who are already experts or leaders in their fields, disciplines or hobbies and who want to use this nature-connecting course process to make their skills organic. In this essential improvement and to the benefit of all, these individuals help reverse rather than increase the earth misery catastrophe that our society continues to create.

If you think you are interested in an 18 month, subsidized, online, GreenWave Organic Therapy degree call 360-378-6313 for further information or visit: http://www.projectnatureconnect.net

Most of what you will eventually need to know or learn for your degree is the heart of this book. It is Industrial Society's missing link to peace on Earth through harmony with Earth.

EARTH MISERY UPDATE

Have you ever had a rewarding childhood or adult experience in a natural area? If so, then you naturally revere this book's contribution to well-being. This is because that nature-connection helped you consciously enjoy part of your inherent, natural, unadulterated self and you would not want to have the joy of that love torn from your life. The misery of that injurious separation would produce conflicts, insensitivities, and disorders in and around us. It would go against your established human rights to life. It would not be legal, ethical or moral.

Unbelievably this nature-disconnection tragedy is increasingly eviscerating life today on personal and global levels. Our established thinking claims that the source of this disaster is unknown and a countermeasure for it is unavailable. That is inaccurate. The organic whole life art and science of revolutionary wisdom enables you and your livelihood to help remedy this dilemma and help others do the same.

Where is your life with regard to the information, above? How does it affect your past and present, your dreams and future?

CHAPTER TEN: Applying GreenWave in a Group

Through the GreenWave-54 Organic Psychology of the Natural Systems Thinking Process in this book, any attraction in nature, backyard or backcountry, that we become conscious of can help us bring Nature's revolutionary wisdom into our lives. A weed growing through the pavement can do it. A park or a potted plant can do it. For example, I participated in a rushed, almost stressful training program for people whose differences kept them arguing and in conflict amongst themselves. They had little interest or time to hear an explanation from me of the unifying and healing benefits of the GreenWave-54 reconnecting with nature process, so they omitted it from their agenda. In the midst of this commotion, a young bird flew into the meeting room through the door. It could not find its way out. Without a word, the behind-schedule, quibbling meeting screeched to a halt. Deep natural attraction feelings for life and hope-filled each person for the moment. For ten minutes that frightened, desperate little bird triggered those seventy people to harmoniously, supportively organize and unify with each other to safely help it find its way back home through the entry door.

Apparently, neither a teacher, preacher or politician was present to educate these leaders about the value of the bird's life and supporting it. Although it said not a word, the bird, nature itself, was that educator. Non-verbal sensory attraction factors within the integrity of its life touched these same influences in the lives of the leaders. That contact sparked into their consciousness, their innate natural feelings of love, in the form of nurturing, empathy, community, friendship, power, humility, reasoning, place, time and many more of our internal 54 natural senses. That reconnecting moment with nature engaged and nourished a battery of their natural sensitivities and sensibilities. These inborn natural intelligences led a contentious group to support rather than deny the life of a bird and to bring new joy to their personal and collective lives for a short period. When they accomplished this feat, **they cheered their role and the human spirit, not the role of the bird.** They felt like heroes for the moment for setting it free and congratulated their humanity for its wisdom and compassion in this effort. In their story of the incident, the role and impact of the bird went unnoticed. They returned to the hubbub of the meeting as if nothing unusual had happened. They completely overlooked that the bird had united them while it was there, something they continued not to do without its presence.

I wanted to say something about the effect of the bird, but I didn't. People would have scoffed. They would have said that what happened was not significant or useful for it was uncommon to have a wild bird interrupt their lives. It was their "human spirit" that they applauded, not its origin and existence in nature even though several Native American groups were part of the training program.

Most people can see that reconnecting with nature during that incident brought a distinct joy and integrity to their lives. The individual and collective benefits were evident. It is the continual lack of such natural attraction contacts that creates our disorders. People feel distraught, yet helpless, about Earth's life and their lives being at risk, like the bird. If you use this book's GreenWave Natural Systems Thinking Process, even an aquarium or pet can produce the same benefits. I've seen it unify couples, families, and individuals in conflict with themselves or others. It is an organic, unifying CRL tool.

We produce and suffer Earth misery because we excessively attach ourselves to exploiting nature, so we lose the integrity of its benefits. Did Parts 1-4 help you learn how to reunite with nature to regain them? That is the heart of the matter that GreenWave-54 addresses.

Using the Zipper Tool would have been appropriate for the meeting because it utilizes real contact with nature, in backyards, parks, even with potted plants, a bird, and wilderness, too. There was a park near the meeting site where we could do, own and teach simple nature-reconnecting activities. They could have shared their conscious sensory contacts with attractions in authentic nature to elicit an essence of cooperative life relationships. This would have enhanced the meeting's goals for unification.

Like playing hide and seek, the GreenWave activities are exciting and interesting. They provide, at will, the nature-reconnected moments missing from our lives. People can make them happen anytime because they know the "rules" of each activity. The process is uplifting and responsible. It helps us nurture as many as 54 natural senses we genetically inherit and share to bring about homeostatic consensus within and around us.

#20 An Expedition Challenge:

Place one of your hands in a bowl of hot water while their other hand is placed in a bowl of very cold water. After a minute, plunge both hands into one bowl of lukewarm water. At that moment, the hand coming from the cold water signals your consciousness that the lukewarm water is hot, while the hand is coming from the hot water signals that the lukewarm water is cold. Senses are not making sense, for time subconsciously adjusts our sensitivity to environmental variations. The sense of reasoning is frustrated and shifts into curiosity and wonder.

Without conscious knowledge of the events leading to this activity's sensory discrepancy, a person can learn to mistrust their natural sense of temperature along with their self-confidence, or falsely think water has miraculous properties. Through new moments of time, the addition of other sensory input from accurate information and reasoning makes a person aware of all the circumstances. We learn something about our sensory nature and learning. This demonstrates the importance of knowing in 54 multisensory ways, yet we excessively discriminate and disregard the sensory information available from actual sensory contact with nature.

Concerning how we relate to the world, how differently we know it if we are brought up connected to nature versus disconnected from nature. Our disconnections presently have us over our heads in hot water[5].

1. Find a safe, attractive area and spend 3 minutes getting to many sense know it with your eyes open.
2. Shut your eyes and spend 3 minutes getting to know it with your eyes closed.
3. What differences did you find between 1 and 2?
4. If you were blind for years and then able to see, what temporary confusions might you experience when visiting this same natural area?
5. How does this apply to the reality of SUNNEH-NNIAAL-54 in our cultural life when 99 percent of our sensing and feeling time is disconnected from it as a natural area?

EXPLORE NOW: VALIDATE. *CONNECT THE CONCEPT, ABOVE, WITH A CONSENTING NATURAL AREA.*

- **SEVMRATCI** the most interesting felt-sense attractions and senses that you find or that find you[40].(Appendix A: Our Fifty-Four Natural Senses and Sensitivities)
- Add this process to an experience you have had in your special area of interest: Art, Creative Writing, Music, Yoga, Parenting, Recovery, Addiction, Renewal, etc.

- **Supportive Reading and Activities:** in a natural area read the next chapter in *Reconnecting with Nature* (or approved course book) including doing its activities and journaling them.

- **Journal Response Form:** For personal growth and reference to describe the value of this section, please complete the Appendix B: Revolutionary Wisdom Response Form: Journal Response Form[41].

Applying GreenWave-54 with an Individual

John Scull, Ph.D. Neuropsychologist said: "GreenWave-54 is one of the methods whereby we can give people immediate reward experiences with environmentally healthy behavior. I have found, in leading Project NatureConnect events that, just as you say, it builds a sense of community, and mutual support. Perhaps most important, it helps people learn to trust their senses and trust in the natural world."

Mike Cohen says we can do some of his activities with a potted plant, and I had one client where this was shown to be correct.

My client had been in a motor vehicle accident and had multiple injuries that prevented her from getting out, and about. She was in chronic pain and was having severe anxiety attacks.

There was a lovely potted plant in her living room (I think it was a dieffenbachia, but I'm not sure). I described how we are in a mutually-supportive role with plants. They provide us with oxygen, and we provide them with CO_2. I suggested that when she was feeling anxious, she should breathe with the plant, exchanging gases and recognize that they were mutually supportive. She did so; I could see her visibly relaxing. Within a few minutes, she reported that her anxiety was gone and her pain was somewhat reduced.

Over the next few weeks, she successfully used this technique whenever she was anxious. As her physical abilities improved, she began spending time in the garden and then began gardening from her wheelchair. Her anxiety was completely controlled without medication.

I don't know what she bonded to or with, but she developed a sense that the plant (and later her garden) was her friend and that they were in a mutually supportive relationship. This feeling was able to help her overcome her anxiety."

EXPLORE NOW: VALIDATE. *CONNECT THE CONCEPT, ABOVE, WITH A CONSENTING NATURAL AREA.*

- **SEVMRATCI** the most interesting felt-sense attractions and senses that you find or that find you[40].(Appendix A: Our Fifty-Four Natural Senses and Sensitivities)

• Add this process to an experience you have had in your special area of interest: Art, Creative Writing, Music, Yoga, Parenting, Recovery, Addiction, Renewal, etc.
• **Supportive Reading and Activities:** in a natural area read the next chapter in *Reconnecting with Nature* (or approved course book) including doing its activities and journaling them.
• **Journal Response Form:** For personal growth and reference to describe the value of this section, please complete the Appendix B: Revolutionary Wisdom Response Form: Journal Response Form[41].

#21 An Expedition Challenge:

A. What attractive senses and feelings[27] do I experience in the life of this area?

B. Ask an attraction, "Who are you without labels and names?" After it responds ask it, or ask yourself, "Who am I without labels and names?"

C. List the differences and similarities in the responses you discovered in B

D. Ask yourself "What would I sense and feel if my ability to register this attraction connection was taken away from me?

Enhance your knowledge. Experience the self-evident facts at http://ww.ecopsych.com/54nineleg.html

Additional Information:
http://www.ecopsych.com/truthlist.html

A Potential final exam question: *Do labels change how you know Nature? (who, where, how, what, when)*

EXPLORE NOW: VALIDATE. *CONNECT THE CONCEPT, ABOVE, WITH A CONSENTING NATURAL AREA.*
• **SEVMRATCI** the most interesting felt-sense attractions and senses that you find or that find you[40].(Appendix A: Our Fifty-Four Natural Senses and Sensitivities)

• Add this process to an experience you have had in your special area of interest: Art, Creative Writing, Music, Yoga, Parenting, Recovery, Addiction, Renewal, etc.
• **Supportive Reading and Activities:** in a natural area read the next chapter in *Reconnecting with Nature* (or approved course book) including doing its activities and journaling them.
• **Journal Response Form:** For personal growth and reference to describe the value of this section, please complete the Appendix B: Revolutionary Wisdom Response Form: Journal Response Form[41].

The next chapter describes the process and effects of taking our typical excessively nature-disconnected lives on a year-long, evidence-based, nature-reconnection expedition that helps its members deal with and reverse the Earth misery that we all suffer one way or another.

The expedition was an objective search for Mother Earth. For ten months it brought our old stories, injuries, feelings, and memories into the pure science experiences of our 54-senses in the GreenWave present while camping out in 84 different natural habitats from Newfoundland to California. Much of the art and science of Organic Psychology was discovered on it.

You can to some extent participate in the year-long expedition as you read about it, by inserting the **EXPLORE NOW: VALIDATE** series, above, to attractive highlights you find in **Chapter 11**, below, as well as discover and strengthen more connections with your companion natural area.

Where is your life with regard to the information, above? How does it affect your past and present, your dreams and future?

CHAPTER ELEVEN: An Expedition Process Summary

Revolutionary wisdom continues a successful, moment-by-moment, learn by doing exploration process that was established in 1965, and that is described here. As it is doing right now, the book and path make you aware of itself. At this point, it says that

- It is 54-sense self-evident that globally and personally we are suffering today's earth misery on many fronts due to our society's excessive disconnection from the life of earth's eons of loving natural relationships in and around us.
- The core of most problems personal, international or global is our 54-sense emotional attachments to stories that conflict with each other. We may be able to understand

what a contradictory story says, but if we are attached to a different story, it does not change, and the new story can push our buttons when we hear it, or it is applied to us.

- This situation does not change until a unifying factor comes into play or one side of the conflict is conquered/removed.
- Reasonable and caring people need to use the art of organic GreenWave-54 science as a unifying tool to help them stop this catastrophe because its omission causes it.
- Question yourself. Is it self-evident to you that you are a caring individual and that you want to continue this learning-by-doing expedition path? If so, we encourage you to proceed and call us if you need further information. 360-378-6313

Multi-Sensory Expedition Education Outcomes

The Source and History of Organic Psychology and GreenWave CRL relationships.

The whole life art and science of Organic Psychology has its unique sensory source in my life experience and training. It is as if Eleanore Roosevelt, Louis Mumford and Felix Adler purposely loved me to be born in Sunnyside Gardens[57], a unique planned, garden community, now historic site, that they built in the New York City Borough of Queens in 1929 because, as Eleanore Roosevelt said, "People lose their humanity when they are too separated from nature".

In Sunnyside Gardens, as well as through my parents and several progressive summer camps[58], **I unknowingly became aware that my left-handedness was part of my nature.** I rudely discovered this when, at age six, I felt my elementary school unreasonably manipulating what I sensed when they demanded that I write right-handed. It disturbed me. It felt abusive to my natural integrity, and it took four years of discomfort and arguments to finally have the school consent to let me write with my left hand using a, then taboo, fountain pen. That nature-supportive use of technology changed and guided my life.

As part of my left-handed challenge, I learned from Nature in and around me that **our society's outdated ways were prejudiced against my Nature** as I discovered that left-handedness was sinister or evil[59] and I could be treated accordingly. I also found that this prejudice could be changed through extended contact with and education about Nature's value and integrity. However, I learned this by being the Guinea Pig in making this change happen at the school, and that experience has made it remain a central focus of my life. For this reason, as new evidence-

---

[57] http://www.millermicro.com/sunnyside.html

[58] http://www.ecopsych.com/history.html

[59] https://en.wikipedia.org/wiki/Bias_against_left-handed_people

based information and technology are discovered that improves our relationship with the life of our planet, I incorporate it in my personal and professional activities, sometimes to the inconvenience of my friends and myself.

A Potential final exam question: *What examples can you find of contemporary thinking being prejudiced against Nature? (who, where, how, what, when)*

EXPLORE NOW: VALIDATE. *CONNECT THE CONCEPT, ABOVE, WITH A CONSENTING NATURAL AREA.*
• **SEVMRATCI** the most interesting felt-sense attractions and senses that you find or that find you[40].(Appendix A: Our Fifty-Four Natural Senses and Sensitivities) • Add this process to an experience you have had in your special area of interest: Art, Creative Writing, Music, Yoga, Parenting, Recovery, Addiction, Renewal, etc.
• **Supportive Reading and Activities:** in a natural area read the next chapter in *Reconnecting with Nature* (or approved course book) including doing its activities and journaling them.
• **Journal Response Form:** For personal growth and reference to describe the value of this section, please complete the Appendix B: Revolutionary Wisdom Response Form: Journal Response Form[41].

Expedition Process: Earth Alive

In 1965 an unusual thunderstorm in the Grand Canyon convinced my extended sensory contact with natural areas that the life of Planet Earth had to be an immense organism because there was nothing I did that it didn't do and vice-versa, except talk. It was like other people, and I lived in it and as part of it as cells that had the unique ability to speak and relate to stories rather than merely live out the wisdom of Earth's life. That insight might let the value of my left-handedness speak for itself as well as for others whose lefty or other trespasses had been admonished by our society's established ways. It felt like I had found another, more loving society and I developed a fiduciary relationship with it. For this reason, I designed and established a learning community of twenty folks. It made space for individuals who would dedicate themselves to together camp out on a nature-connecting expedition across the USA for a year. The purpose of their exploration was to discover if they could produce a fiduciary, consensual interaction, a learn-by-doing process that was in the balance with the natural world. Each year each new group would reinvent the expedition process so, as it succeeded, each participant might know how to best live their life harmoniously with the life of Nature in the environment, each other and themselves.

A Potential final exam question: *Do you know of any evidence that shows Earth is not alive? (who, where, how, what, when)*

EXPLORE NOW: VALIDATE. *CONNECT THE CONCEPT, ABOVE, WITH A CONSENTING NATURAL AREA.*
• **SEVMRATCI** the most interesting felt-sense attractions and senses that you find or that find you[40].(Appendix A: Our Fifty-Four Natural Senses and Sensitivities) • Add this process to an experience you have had in your special area of interest: Art, Creative Writing, Music, Yoga, Parenting, Recovery, Addiction, Renewal, etc.
• **Supportive Reading and Activities:** in a natural area read the next chapter in *Reconnecting with Nature* (or approved course book) including doing its activities and journaling them.
• **Journal Response Form:** For personal growth and reference to describe the value of this section, please complete the Appendix B: Revolutionary Wisdom Response Form: Journal Response Form[41].

Guided by my living earth perceptions, commencing in 1968, for sixteen years every September a yellow school bus and its occupants embarked on a ten-month living and learning expedition into natural areas. Outfitted only with camping gear and a small library, it carried twenty students, my staff and myself into a personal and academic growth utopia. It began my sleeping out under storms and bright stars year-round; camping exploring and learning while enveloped in America's spectacular natural areas.

This all-season, consensus-governed, outdoor-living program that I founded in 1959 immersed its friendly school community in rich organic critical thinking, interpersonal experiences and natural wonders. Participants thrived in eighty-three different natural habitats, and from keeping their commitments to open, honest relationships with the environment, each other and with indigenous people(s), researchers, ecologists, the Amish, farmers, anthropologists, folk musicians, naturalists, shamans, administrators, historians, counselors, teachers and many others. The experience profoundly connected our inner nature to the whole of nature.

Because of our romance with educating ourselves this way, in the school community:

- Chemical dependencies, including alcohol and tobacco, disappeared as did destructive interpersonal and social relationships.
- Personality and eating disorders subsided.
- Violence, crime, and prejudice were unknown in the group.
- Academics improved because they were applicable, hands-on, and fun.

- Loneliness, hostility, and depression subsided.
- Group interactions allowed for stress release and management; each day was fulfilling and relatively peaceful.
- Senses that had been injured in childhood were recognized as being vulnerable to individuals who, due to similar childhood injuries, were less sensitive to other's vulnerability to these senses. Connections with natural areas replaced these vulnerabilities in both parties and increased unity.
- Students using meditation found they no longer needed to use it to feel whole because an extra organic dimension was added to the process.
- Participants knew each other better than they knew their families or best friends. They risked expressing and acting on their more profound thoughts and feelings; a deep, fiduciary sense of social and environmental responsibility guided their decisions.
- When vacation periods arrived, only a few wanted to go home. Each person enjoyably worked to build this supportive, balanced living and learning utopia. They were home.

All this occurred purely because every community member met their commitment to making sense of their lives by establishing relationships that supported the expedition and the natural world within and around them. If we didn't, we would be unable to operate. We hunted, gathered and practiced such relationships; we organized and preserved group living processes that awakened our natural wisdom. We explored how to let Nature help us regenerate responsible connections when they decayed.

EXPLORE NOW: VALIDATE. *CONNECT THE CONCEPT, ABOVE, WITH A CONSENTING NATURAL AREA.*
• **SEVMRATCI** the most interesting felt-sense attractions and senses that you find or that find you[40].(Appendix A: Our Fifty-Four Natural Senses and Sensitivities) • Add this process to an experience you have had in your special area of interest: Art, Creative Writing, Music, Yoga, Parenting, Recovery, Addiction, Renewal, etc.
• **Supportive Reading and Activities:** in a natural area read the next chapter in *Reconnecting with Nature* (or approved course book) including doing its activities and journaling them.
• **Journal Response Form:** For personal growth and reference to describe the value of this section, please complete the Appendix B: Revolutionary Wisdom Response Form: Journal Response Form[41].

The secret to our expedition's success was to learn to tap into and learn directly from the natural world, the life of earth within and about us. Through our 54 natural attraction sensations and feelings it taught our sense of reason how to trust the life of Nature and how to validate and incorporate its reasonableness. The highlights of this process included:

- **Starting the school year in a classroom that had no individual or authority sitting in the front of the room desk.** The staff, unrecognized by some, sat with the students facing the desk and blackboards. In time students and faculty started to ask each other what was happening and introduced themselves to each other. As the discussion proceeded, the group rearranged the desk chairs so they were now in a circle and folks could better relate to each other. **That moment was when we started the school year.**
- All aspects of the program except driving the school bus and some administrative duties were self-organized and shared by all. This included governing by consensus. Anybody could call an all-school meeting at any time to address challenges or discontents that arose. This process became the core of the curriculum.
- The daily and all-year itinerary were reasonably developed based on weather conditions, distinct connections, and geographic logistics. It included spontaneous opportunities or challenges as they arose and it could be changed by consensus as new ideas and information appeared.
- By agreement technological assistance and dependence was limited so that the community depended upon nature and each other for information and satisfactions. Technologies were limited to the bus, camping equipment, cameras and a small library, no radios, TVs or tape players. The only media player allowed was a tape recorder that played back to us interviews, songs, music, and reactions to and from our expedition experiences.
- The group was a rational combination of Kurt Lewin T-group encounter processes, Progressive Education, and Alexander Wolf group therapy. The latter recognized that group members subconsciously registered a group they were in as their family at home and often felt the joy or hurt of the latter and acted accordingly in the expedition group. In our nature-connected family, we learned from experience how to express CRL and guide our relationships with using nature's 54-sense wisdom. We let CRL become the boss of us.
- When expedition members found that some of their senses were injured or weak, they discovered that it made sense to temporarily ride on the senses and feelings of others who were healthy in these same areas. This continually made the expedition group a unified whole and helped individuals restore their balance through bonding connections with attractive natural senses of individuals in the group and attractions in natural areas that surrounded us. An expedition was organized so that in its membership a wide range of injured sense challenges supported and educated it and repaired them.

- In reality and imagination each member of the community, including its staff, was roped into each other's personal lives for the year as would be a mountain climbing expedition that organized and maintained itself to survive by helping its environment survive. This enabled us to sense and feel our actual relationship with the life of Earth, and it's web members because, without today's "ticket to Mars," we are each on an expedition to support and enjoy the life we share with our planet home and each other.
- Our special high motivation survival goal was consistently re-affirmed: this was a one-year expedition that could dissolve if it did not maintain itself appropriately. It was also the long-term goal and livelihood of some group members that went far beyond one year's duration, (For myself it was sixteen years).

Again, it is significant that in a short period of time, globally, within local areas, this same organic, small-group expedition education process can be included in the eleven sections that the PNC mission addresses (page 42). Also, these local groups can globally learn from each other via the internet.

The National Audubon Society identified the Trailside Country School expedition education program as **the most revolutionary form of learning in America** and selected it to be the official Audubon alternative education program out of 116 programs that were considered. It became the National Audubon Society Expedition Institute in 1978.

EXPLORE NOW: VALIDATE. *CONNECT THE CONCEPT, ABOVE, WITH A CONSENTING NATURAL AREA.*
• **SEVMRATCI** the most interesting felt-sense attractions and senses that you find or that find you[40].(Appendix A: Our Fifty-Four Natural Senses and Sensitivities) • Add this process to an experience you have had in your special area of interest: Art, Creative Writing, Music, Yoga, Parenting, Recovery, Addiction, Renewal, etc.
• **Supportive Reading and Activities:** in a natural area read the next chapter in *Reconnecting with Nature* (or approved course book) including doing its activities and journaling them.
• **Journal Response Form:** For personal growth and reference to describe the value of this section, please complete the Appendix B: Revolutionary Wisdom Response Form: Journal Response Form[41].

EXPEDITION GREENWAVE CRL ORGANICS

Starting in 1960, from thirty years of all-season travel and study in over two hundred sixty national parks, forests and sub-cultures, I developed a learning process and psychology that unleashed our natural ability to grow and survive responsibly. By documenting that it worked repeatedly and could be taught, I earned my doctoral degree and the school became a small, accredited graduate and undergraduate degree program at several Universities. I have helped that experience further grow and develop since 1990, by, for 26 years daily climbing a local wild mountain overlooking the sea (that is 27,000 miles, 62,000 hours) and sleeping outdoors, year-round, since 1969.

To share my discovery with the public, in 1984 I initiated the National Audubon Society International Symposium "Is the Earth a Living Organism?" and in 1987 I encapsulated my nature-connected psychology as Project NatureConnect in a series of 154 sensory backyard and backcountry activities (Cohen, 1988, 1990, 1993). They have been used as an organic form of psychology in many courses and training programs for educators, parents, and students so the latter may incorporate these revolutionary interactions in their work and lives. These simple, fun nature explorations benefit people of all ages and backgrounds. Uniquely, they revitalize the innate sensory communication and support between a person and the natural world, in and around others and themselves.

Organic Psychology natural area activities balance our lives by letting our 54-sense natural connectedness identify and be our common cause of survival in balance. By reducing stress while motivating participation, the activities promote recovery from destructive habits, dependencies, and dysfunctions. Today professionals use them to augment counseling, twelve-step, hospice, stress management, conflict resolution, self-esteem, and environmental education programs (Cohen, 1993). They help us follow the suggestion in Job: "Speak to the Earth, and it will teach thee." They go more in-depth and more sanely than most psychological and psychiatric practices.

Because Organic Psychology honors the wisdom of the global life community, it enables that community to relate and grow harmoniously through the natural attractions of its attraction-field root. That fabric, which we each biologically inherit, lovingly unifies life relationships rather than further subdividing Earth into separate academic and institutional cubbyholes. My extensive time engaged this way enables me at will to snuff out my story way of knowing and thoroughly enjoy unity with the natural intelligence that loved me into being.

In reality and imagination, the art and science of organic psychology expeditions return us to our origins. There we critically measure information, procedures, and behaviors by their natural attractiveness and long-term effects. We then, moment by moment, responsibly organize and balance our relationships by helping our natural attractions unite themselves.

EXPLORE NOW: VALIDATE. *CONNECT THE CONCEPT, ABOVE, WITH A CONSENTING NATURAL AREA.*
• **SEVMRATCI** the most interesting felt-sense attractions and senses that you find or that find you[40].(Appendix A: Our Fifty-Four Natural Senses and Sensitivities) • Add this process to an experience you have had in your special area of interest: Art, Creative Writing, Music, Yoga, Parenting, Recovery, Addiction, Renewal, etc.
• **Supportive Reading and Activities:** in a natural area read the next chapter in *Reconnecting with Nature* (or approved course book) including doing its activities and journaling them.
• **Journal Response Form:** For personal growth and reference to describe the value of this section, please complete the Appendix B: Revolutionary Wisdom Response Form: Journal Response Form[41].

Various Labels

For sixty years I have pioneered and promoted the scientific value of creating moments that let Earth teach us what we need to enjoy the wisdom of nature's garbage-free perfection. This has had various names assigned to the process as information advanced including:

- 1959 Trailside County School outdoor travel and camping
- 1977 An Expedition Education learning program via the Regents of University of the State of New York
- 1965 Organism Earth: the life of Earth having its own, self-correcting homeostasis
- 1978 National Audubon Society Expedition Institute at Lesley University
- 1986 Integrated Ecology at World Peace University
- 1990 Project NatureConnect in cooperation with Portland State University
- 1990 NSTP: The web-of-life Natural System Thinking Process
- 1990 Nature-connected 54-sense knowing and relationships,
- 1991 Integrated Natural Attraction Ecology at the Institute of Global Education
- 2005 Akamai University Institute of Applied Ecopsychology
- 2013 The Organic Psychology GreenWave-54 Albert Einstein Unified Field process and the equation for Eco-Art, and Organic Therapies.

These years and results testify to my stubborn dedication to Nature's welfare and that I experience it as my peculiar integrity. I thank my parents and Eleanor Roosevelt for this precious knowledge along with the exceptional individuals who continue to participate in and support Project NatureConnect.

EcoArt Activity: GreenWave: Opening Your Heart to the Darkness of the Night

#22 An Expedition Challenge:

Living and learning on the Expedition was in some ways as different as night and day in comparison to "normal" life. Does this activity help you bridge the gap?

As an adult, I have spent many nights looking at the stars, moon, and Milky Way. These experiences express themselves very differently than they did when I lived in the city. The absence of humanmade light and sounds can be an incredible experience. I have learned how to allow my heart to open to the living world just as the darkness opens itself to the starry night and moon. I have encountered moths, June bugs, fireflies, owls, wolves, coyotes, mountain lion, deer, katydids, and so much more!

Your love of whole life in action.

Our senses register our awareness of the world that is in and around us. Our thoughts, feelings, and relationships are built on self-evident information that our senses convey. When the evidence or our senses are limited, warped or injured, so are our lives, awareness, and happiness.

Whenever you find an attraction in a natural area or the nature of another person engage and validate your senses of Consciousness, **R**eason, and **L**anguage **(CRL)** to powerfully blend and express themselves in an informed way to others who might appreciate and grow from them.

1. At night, choose an area that is attractive. Keep in mind that you want to experience the night time living systems as natural as possible. For example, instead of sitting near a humanmade light try to take a few steps away from this light so you can emerge yourself safely in the pureness of the dark. Remember to **Gain consent to visit** a natural area, including a person's inner nature.
2. Find an attraction(s) there or give it time to find you.
3. **From the 54-sense list** say or write the name of the self-evident senses, sensations or sensibilities that you register. Do this as well for other attractions that show up in your immediate experience.

4. From your experiences draw what your attractions are in the nighttime. Remember you may not be able to directly see what a living being is so you can be creative and draw what you think the living being looks like. Try not to draw from memory or images. Let your inner artistic nature guide you.
5. Title and sign your artwork.

Reflective Questions

1. How hard or easy was it for you to connect with the natural environment versus the human-made environment?
2. What noises in the human-made environment did you have to block out, or try to, in your experience?
3. Did you experience nature differently at night than you do in the daytime? If so how? What new senses/sensations did you experience?
4. Did you experience cause you to feel negative in any way? If so how and how did you self-correct this?
5. Are you happy with your artwork? Why or Why not?
6. Did it feel natural to participate in nature connecting experiences at night? Why or Why not?
7. What did you learn from this activity?
8. What did you like the most about this activity?
9. What did you like least about this activity?
10. Do you have suggestions for improving or adding to this activity?

EXPLORE NOW: VALIDATE. *CONNECT THE CONCEPT, ABOVE, WITH A CONSENTING NATURAL AREA.*
• **SEVMRATCI** the most interesting felt-sense attractions and senses that you find or that find you[40].(Appendix A: Our Fifty-Four Natural Senses and Sensitivities) • Add this process to an experience you have had in your special area of interest: Art, Creative Writing, Music, Yoga, Parenting, Recovery, Addiction, Renewal, etc.
• **Supportive Reading and Activities:** in a natural area read the next chapter in *Reconnecting with Nature* (or approved course book) including doing its activities and journaling them.
• **Journal Response Form:** For personal growth and reference to describe the value of this section, please complete the Appendix B: Revolutionary Wisdom Response Form: Journal Response Form[41].

THE WARRANTIED FACT CHECK: Visit the warrantied fact list http://www.ecopsych.com/54warrantfact.docx and insert dates on new facts that you learned or know to this point in the book. Place a check mark on facts you previously dated that you feel have been reinforced or extended. Where is your life with regard to the information, above? How does it affect your past and present, your dreams and future?

PART SIX

The Natural Systems Thinking Process (NSTP) Web-of-Life Model

"I am especially awed by the integrity and creative genius of designing this online experience that has the potential to reach so many people. This demonstrates responsible use of technology. I learned to trust myself and that it was okay to trust myself.

While I have always enjoyed experiences in Nature, there was a subtle tension surrounding them, something that would not allow me to 'feel' them completely or acknowledge them openly. During the course activities and readings, I began to understand why and how that came to be. That, in addition to sharing with our group, freed me to acknowledge, validate and continue to seek similar enjoyable experiences: Life-changing.... life-affirming.... a blessing to share.... a gift to receive.

I am worth more than I realized. I can go anywhere and be connected to something huge that is part of us all. The course helps me fill the void that most of us have inside. It shows us how society has tainted our thoughts. It offers so much potential to heal mother earth and all her life. If all people knew how to connect, there would be no more violation of life. I have been using some of the techniques and incorporating it into my art classes with at-risk young students. The response has been wonderful."

Discovery of the Higgs Boson

With the discovery of the Big Bang Higgs Boson natural attraction field announced in July 2012 after eighty years of research, the source of our life-support deterioration along with the remedy for it further affirmed the work of Project NatureConnect (PNC) and its ***Natural Systems Thinking Process*** **(NSTP)** Web of Life Model. This was because the core of PNC had historically been that natural attractions or attachments are the GreenWave strands and webstrings that make up every aspect of the web-of-life. Fifty years ago, that model postulated that attraction was the core of Nature and the life of Planet Earth's web of life acted accordingly.

The GreenWave power and wisdom of NSTP is that scientifically the Universe, Nature, Earth, and Humanity are moment-by-moment all-attraction expressions and manifestations of the Unified Field. **Nothing is unattractive or negative in or around us except the nature-disconnected stories that we attach to our 54 GreenWave senses.** The way we think with these stories deny or disrupt the all-attraction truth and value of NSTP whole-life science.

> *For example,* A rubber balloon filled with a mixture of gases including nitrogen, oxygen and carbon dioxide has its attractive wholeness.
>
> When you push in its rubber skin, you stressfully separate the skin molecules by stretching them out. At the same time, you attract together the air molecules within the balloon to accommodate the push and its value and their crowded attachments stress them.
>
> Both the separation and attraction are "dance properties" of the balloon that your push energizes into dancing. Both make possible an attractive pulse that supports the balloon's whole-life survival in coping with the change from a push. Neither are negatives.
>
> If over time, your story that pushes the balloon is not attracted to more attractive things in response to the stressed resistance dance signals it gets from the balloon; your story is then a negative because it injuriously stresses the balloon. It is not part of how the balloon typically operates to sustain its well-being by its energies bringing themselves into balance.

When the balloon is the web of life of Planet Earth, the consequence of our addictive and increasingly pushy, planet-insensitive stories is Earth Misery.

Presently, **over ninety-nine percent of the stories that we use** to think, feel and relate typically are disconnected from the way the life of Earth.

The Project NatureConnect NSTP model and its stories are excellent for learning how Earth works because strands of the web represent attracted, connected, sensitive Web of Life GreenWave connections that include 54 natural sense signals from the life pulse of Earth.

EXPLORE NOW: VALIDATE. *CONNECT THE CONCEPT, ABOVE, WITH A CONSENTING NATURAL AREA.*

• **SEVMRATCI** the most interesting felt-sense attractions and senses that you find or that find you[40]**.**(Appendix A: Our Fifty-Four Natural Senses and Sensitivities) • Add this process to an experience you have had in your special area of interest: Art, Creative Writing, Music, Yoga, Parenting, Recovery, Addiction, Renewal, etc.
• **Supportive Reading and Activities:** in a natural area read the next chapter in *Reconnecting with Nature* (or approved course book) including doing its activities and journaling them.
• **Journal Response Form:** For personal growth and reference to describe the value of this section, please complete the Appendix B: Revolutionary Wisdom Response Form: Journal Response Form[41].

The Webstring Natural Attraction Model (NSTP)

Activities for this model are done in conjunction with an online group http://www.ecopsych.com/orient.html that starts every few weeks. Write or call for information nature@interisland.net or 360-378-6313.

For almost sixty years, Michael J. Cohen[60], the creator of NSTP, has been an environmental educator and psychologist who, from many decades of university training and living and learning in natural areas, became aware that our socialization in an industrial society prejudiced our mind. **It conditioned us to know nature and ourselves through abstract stories that separated our thinking from the beauty and balanced ways of nature, and this created our disorders.** The separation biased the way we think to believe that it alone was intelligent and that our thinking made more sense than the natural attraction way that nature worked to produce its eons of self-correcting perfection.

Upon completion of his undergraduate and graduate studies in natural science and counseling in 1957, for 49 years Cohen increasingly lived, learned and researched ecologically sound relationships as he guided expedition education camping and study groups into natural areas for periods of thirty days to a year at a time. From this remarkable outdoor experience, **he recognized that he and Planet Earth were equally alive and shared all aspects of life except one, humanity could build relationships using stories, the life of Earth could not.** Over time, he developed a sensory nature-connecting model that empowered its participants to genuinely connect their thinking and relationships to the balance and renewing powers of nature, the real thing, backyard or backcountry.

---

[60] http://www.ecopsych.com/mjcohen.html

Cohen's unique Webstring Natural Attraction Model (NSTP) enabled the thinking of its participants to sensibly become familiar with, respect and enhance nature's nurturing life-flow in and around them. It benefited them by giving them the means to eliminate the separation of the life of their psyche from nature's restorative intelligence and balance. He observed that it was this separation that made industrial society reduce personal, social and environmental well-being.

Today, the NSTP model's process helps us transform our hurtful ignorance and deterioration of the web of life into mutually supportive relationships with its natural systems in the environment, other people and ourselves. Also, the Model gives us the means to teach others how to accomplish this quickly.

Because the Webstring Model has successfully involved people in the process of reducing their destructive relationships with nature, the Model is significant because it is far more experiential and practical than theoretical. It provides us with the means to achieve our most valued goals. This is important because we best increase well-being by owning and using tools that help us build mutually supportive relationships with nature that reduce our dysfunctions.

By 1948, history and current events demonstrated that our thinking in industrial society was programmed to conquer, exploit and control nature for profit, not to embrace nature. Embracing nature was considered, flakey, subjective, unscientific, touchy-feely tree-hugging. We not only learned to think that loving nature impeded "progress" and "economic growth," we were paid and in other ways rewarded to think this way. This was no small matter nor was it a secret. It was a matter of massive and long-term public consciousness that industrial society had bonded to support our harmful ways, even when we hated their personal and global adverse effects. By also learning to ignore appropriate tools to deal with this phenomenon, most people were rendered helpless and apathetic in this regard. They knew that the nature-destructive thinking and extreme ways of industrial society deteriorated natural systems within and around them. They had yet to recognize the source of this dilemma and how to remedy it via NSTP.

The Process of the Webstring Natural Attraction Model

In his classic 1953 book, *The Web of Life,* a first book of Ecology, John Storer, incorporating Eugene Odum's scientific methodology in *Fundamentals of Ecology,* brought to public attention the fundamental truth that all aspects of life are related to each other and that this gives life the ability to create its supportive environment and healthy balance. Storer noted that, through Ecology, he described the web of life concerning what could be identified as an orderly progression of significant food chains and energy threads that were only a small part of the massive facts and forces that go into making the physical global life community. Ecology provided scientific evidence for the concept of a universal form of oneness that all things are connected. This had long been part of human thinking in many cultures.

In 1972, two years after the first Earth Day, Cohen watched an environmental education specialist in Smokey Mountain National Park ecologically demonstrate how all threads, not just food and energy threads, of the natural community, fit into a single pattern to connect, grow and sustain the massive web of life that Storer and Odum identified. The specialist went beyond learning from book knowledge and theory alone. She involved her audience in an environmental studies model, in an activity that helped them bring to mind, include and validate their personal and professional life experiences. In a natural area, she engaged them in a web of life ecology activity that enabled people of all ages to understand, model and feel the natural environment so they could more appreciate, support and protect it.

The specialist's activity consisted of placing a group of forty park visitors, including children, in a circle and giving each person a card to wear. On each card, some part of nature was inscribed: bird, soil, water, tree, air, wolf, etc. A big ball of string (the GreenWave) was then used to demonstrate the interconnecting relationships between things in nature. For example, the bird ate insects, so the string was unrolled and passed from the "bird person" through the hands of the "insect person." The string represented their connection. The insect lived in a flower, so the string was further unrolled across the circle through the hands of the "flower person." The soil supported the flower, so the string continued across the circle and through the hands of the “soil person.” In time, the ball of string became a web of strings (webstrings) that passed through the hands of the participants and interconnected all parts of nature with each other. It was a science-based, ecologically correct and environmentally sound educational portrayal of the total global life community, including minerals and energies.

The activity continued by requesting that the group, the web of life participants, gently lean away from the web they built while holding it. They sensed and enjoyed how this thin GreenWave string now peacefully united, supported and interconnected them and all of life. The specialist had them note that the more significant the number of nature-representatives that were in the circle, the stronger the web would become. Some people shared how the web was beautiful or how past contact with nature had been a compelling experience that opened new vistas, renewed or even healed them. Most acknowledged that being in nature reduced their stress, even on just a short walk in the park. Some said that nature was their higher power.

A few participants observed that there was no garbage or pollution in this web of life community, nothing was left out, everything belonged and cooperated, even though many things, like a mouse and a tree, were remarkably different from each other. The activity and its discussion evoked feelings of trust, integrity, and unity amongst the participants along with an enormous respect for nature's peaceful diversity.

Having involved people in a webstring model that captured and conveyed nature's perfection, the specialist then cut one strand of the web signifying the pollution or loss of a species, habitat or relationship. The weakening effect on all was noted, not only physically through the string, but also by sadness that many participants felt. As people shared other environmental and social destruction or pollution that they had witnessed or knew about, another and another string was cut. String by string, the web's integrity, support, and power disintegrated along with its spirit. Because this reflected the reality of their lives, participants, some in tears, said that if they felt hurt, despair and futility while others became angry about the loss.

Webstrings, Real or Imagined?

Sixteen years later, in 1988, at an environmental education conference where the same web of life activity was informally demonstrated, Cohen asked the activity participants if they had ever visited a natural area and had seen strings interconnecting things there. They said, no, that would be crazy. He responded, "If there are no strings there, what then are the actual strands that interconnect and hold the natural community together in balance?" It became very, very quiet. Too quiet. **That silence flagged an important missing fact in contemporary thinking, consciousness, and relationships, a fact that is still missing today.** Participants concluded that the question went beyond the scope of environmental education, environmental studies or science. However, Cohen argued that webstrings were a vital part of the web of life and survival. He said **they were just as real and important as the plants, animals, minerals, and energies that they interconnected, including humanity.** If things were connected, as in the model, then the strings were as true, or truer, than 2 + 2 = 4. They were facts as genuine as trees, thirst or motion, water, sight or sunlight. Without knowing, sensing or respecting the webstrings that make up nature and our inner nature, we broke, injured and ignored them, and part of ourselves as interconnected citizens of this global community. As members of the web of life, we were born with the ability to sense, think with and benefit from the webstrings that connected us to nature and our living planet, Mother Earth.

To identify and explain the strings in the web of life, from his 38-year livelihood living and learning in natural areas Cohen modified the web of life activity. His goal was to help its participants be fully aware of webstrings, what they were and their significance. In his version of the activity, he did not start the demonstration with the labeled cards. Instead, he began it by asking participants to 1) visit the natural area around them for five minutes, 2) find two or three things there that for at least five seconds they felt attracted to, 3) identify what they liked about these attractive natural things and then 4) return to the web of life circle. Upon returning, participants, in turn, wrote on a card that they later wore, one of the natural attractions they found, one that, if possible, had not already been chosen by another person to ensure an optimum of diversity in the circle.

To help people integrate the attractions, they found in nature as part of the web-of-life activity; Cohen explained to the participants that, although they could not notice it, the gigantic ball of string faintly pulsated like a heart because it was the source of conscious attraction. The dance of its pulse resulted from, back and forth, being attracted to survive at the moment and then to attractions that drew it into more supportive survival in the next moment. Also, whenever he passed the string through a person's hand, it was always from left to right.

Significantly, Cohen added two more cards to the activity. Each of them was labeled "person," and after all the webstring connections were made between all the participants, including the two people labeled "person," the two "person" people were additionally connected to each other by a special red ribbon. It lay across the top of all the other webstrings in the circle. Cohen explained that **the ribbon signified that people inherited a sense of literacy, a unique ability to connect with each other by thinking and communicating through words, numbers, and stories.** Most participants agreed that the web of life itself communicated or communed through webstrings, but things in nature did not directly think with, understand or use spoken or written words, numbers or stories. In this sense, nature and the natural were illiterate, and our "red ribbon" stories were foreign to them. Cohen noted that this disconnection was very dangerous because it was self-evident that to be part of a system a thing must be in communication with the system and vice versa. **Many participants displayed alarm or sadness to the point of tears during this procedure.**

Cohen added another step to the original activity. He had participants slowly, with their fingers and free hand, move the strings through their hand from left to right so that the pulsating string moved/flowed throughout the web. This depicted the long-term dance and flow of natural systems and their attraction energies. **It portrayed how the "waste products" of one natural thing were a life-giving contribution to other things, including people.** For example, the carbon dioxide we exhaled was food for plants. The model also demonstrated that whenever any member of the web of life community restricted the flow of the string, the string was stretched and stressed until the member cooperated. This tension modified its pulse. However, if the restriction was not corrected, the flow and well-being of each member, as well as the whole system, was adversely affected. This included a reduction in the well-being of the non-supportive member.

In his enhanced web of life activity, during the part of the activity when the strings were being cut, Cohen suggested that it was inaccurate, incomplete or biased "red ribbon" human thoughts and stories about life and natural relationships that mislead people to excessively cut the webstrings or restrict their flow. Non-supportive stories impacted the strings, and therefore the whole web, as they passed through the hands of people. For example, the story of industrial society socialized people to ignore the attraction string flow in each moment as their source of survival and, instead, for profit, survive by excessively devouring or exploiting other members of the web community. **This occurred even to the point that other web members became extinct and disappeared from the global life community.** Participants admitted that this happened in their lives out of habit, addiction or necessity so they could not stop injuring the strings, even when their sense of reason knew it made sense to stop. For example, they could not walk to work if their workplace was not within walking distance.

Since well before the day that the web of life activity was first designed, **Earth and its people were increasingly suffering from "cut string" disintegration, yet we continued to cut the webstrings** at an alarming rate. Few disputed the accuracy of our situation as depicted by the web of life model yet, sadly, people witnessed and felt, especially since Earth Day, the continued cut-string deterioration of the natural world and their relationship with it, each other, and themselves. To those whose belief system rejected that people were part of the web of life, Cohen asked how they explained that in a study published in the Annual Report for Smithsonian Institution in 1953, scientists found by using radioactive tagged atoms that 98 percent of the atoms of our body and mind are replaced each year by atoms from the environment. **Every seven years or so, practically every molecule in our body returns to the environment and is replaced by a new molecule from the environment, just as the string flow of his webstring model portrayed.** Also, our body consists of ten times more non-human cells and organisms than human, cells. Over 115 species, alone, live on and help sustain the health of our skin. (Margulis & Sagan 1986). Cohen would ask, “Doesn't this suggest that we are part of the web of life and it is part of us? Isn't the web like the womb of our post-natal life?” **Are not the life of Earth and ourselves identical?**

In one of his following workshops, Cohen helped participants answer, “What are the strings?” the question that he originally asked at the conference. He suggested that to answer it, they explore and express what they had been sensing or feeling like part of the web of life because the life of their psyche and mind was also part of the web. To this end, he had participants do an activity to help them discover what webstrings might be: “Find any attractive object or thing in the natural area here, and with your total energy pull or push it, but don't dislodge it from its attachment or move it from its place.” Through this activity, **participants were able to physically sense and feel some of the attraction energies that connected things to each other and the whole of the web of life,** including attractions in air and water. They also became aware that their felt-sense attractions to things in nature were likewise webstrings and that webstrings pulsated. As they changed from moment to moment, they re-registered in the things they connected. For example, if a person saw a bird and the bird saw the person, the bird might move, the person might move in return, and this pulse would continue until other attractions called. Both the bird and person registered and reacted to the in-balance webstring senses of motion sight and distance, and perhaps many others, as well.

In time, participants recognized that webstrings were a dance of connective attractions by all of life to obtain things like food, water, habitat, energy, minerals, warmth, community, and support. Soon they realized that, in their psyche, these attractions were also specific senses or desires that they experienced such as hunger, thirst, trust, belonging, respiration and place. They saw how, in the web of life demonstration, every part of the global life community, from the spaces between sub-atomic particles to weather systems, to the solar system, to the life of their mind was included, as part of the web and that everything consisted of, and was held together by webstring attractions and their aware/conscious contact with of each other. This explained why, like the whole web of strings they had constructed, **nothing ordinarily fell apart or became garbage without because that strengthened the entire web.** Webstrings naturally bonded things together in mutually supportive ways. Natural things acted as if they consented with each other to support their individual lives and all of life. This made Cohen hypothesize that there was a source of the strings in the Universe. (The Higgs Boson discovery in 2012 affirmed this as Sense #54.)

Webstrings injured in childhood are recognized as vulnerable to other individuals who, due to similar childhood injuries, are less sensitive to their vulnerability, and vice versa. This situation triggers distress. Webstring attractions in natural areas replace these vulnerabilities, reduce suffering and promote unity.

Because the Webstring Model has successfully involved people in the process of reducing their destructive relationships with nature, the Model is significant because it is far more experiential and practical than theoretical. It provides us with the means to achieve our most valued goals. This is important because **we best increase well-being by owning and using tools that help us build mutually supportive relationships with nature that reduce our dysfunctions.**

EXPLORE NOW: VALIDATE. *CONNECT THE CONCEPT, ABOVE, WITH A CONSENTING NATURAL AREA.*
• **SEVMRATCI** the most interesting felt-sense attractions and senses that you find or that find you[40].(Appendix A: Our Fifty-Four Natural Senses and Sensitivities) • Add this process to an experience you have had in your special area of interest: Art, Creative Writing, Music, Yoga, Parenting, Recovery, Addiction, Renewal, etc.
• **Supportive Reading and Activities:** in a natural area read the next chapter in *Reconnecting with Nature* (or approved course book) including doing its activities and journaling them.

- **Journal Response Form:** For personal growth and reference to describe the value of this section, please complete the Appendix B: Revolutionary Wisdom Response Form: Journal Response Form[41].

#23 Expedition Challenge Activity

This activity helps you experience a crucial point in my life and yours. It allows the dead to rise from the grave. Once I was able to see life everywhere, I could critically validate the universality of my natural senses and stories that connected me to it.

1. Go to an attractive natural area and thankfully gain permission to spend an hour inventorying what is taking place there. Identify how and where for each thing you find, this same thing is taking place in yourself and affecting you, and vice-versa.

2. Do this same activity with a partner and each of the 54-senses that you can find in that person.

Participants' Reactions have been:

- Plants breathe, I breathe.
- I move, rocks move.
- Water cycles through me and everything else.
- Rocks interact, I interact.
- Crystals grow, I grow.
- Soil nourishes, I nourish.
- Atoms attract and are attracted, me too.
- I reproduce, minerals reproduce.
- Birds fly, I jump.
- I desire to be, so does everything else.
- I love to relate; oxygen loves to relate.
- Water changes, I change.
- Roots need diversity, so do I.
- My life results from and celebrates the lives of other beings.
- The desire to be of trees, rocks, and dinosaurs are alive and well within me.

- I'm alive. Earth is alive. I'm not old enough to reproduce, but that doesn't mean I'm not alive.

EXPLORE NOW: VALIDATE. *CONNECT THE CONCEPT, ABOVE, WITH A CONSENTING NATURAL AREA.*
• **SEVMRATCI** the most interesting felt-sense attractions and senses that you find or that find you[40]. (Appendix A: Our Fifty-Four Natural Senses and Sensitivities) • Add this process to an experience you have had in your special area of interest: Art, Creative Writing, Music, Yoga, Parenting, Recovery, Addiction, Renewal, etc.
• **Supportive Reading and Activities:** in a natural area read the next chapter in *Reconnecting with Nature* (or approved course book) including doing its activities and journaling them.
• **Journal Response Form:** For personal growth and reference to describe the value of this section, please complete the Appendix B: Revolutionary Wisdom Response Form: Journal Response Form[41].

#24 Expedition Challenge Activity

Let SUNEH-NNIAAL help you answer a question or strengthen yourself.

1) Identify a question you want assistance with.

2) Find a natural attraction in a natural area, an attraction that you can touch (tree, brook, flower, place, air, etc.) and gain its consent to do this activity with it.

3) Get to know this tree as yourself by identifying as many of the 54-senses that you share with it as you can in this moment of the time-space Unified Field.

> -Validate that now is when the life of its and your essence are identical,
> -Validate, too, that during the growth of Earth's life through the eons this natural, attraction preceded humanity and part of it loved itself into being to your human life so in real time,
> -Validate that what is attractive to you about this attraction is it is you doing the attracting.

4) Now, while unified with the attraction, continue to put into language what attractions you are experiencing that could be helpful to your question.

5) In your imagination become the attraction talking to you and telling you the invaluable experiences you found in 4).

6) Identify the insights or values you obtained from 1-5 and apply them when appropriate.

EXPLORE NOW: VALIDATE. *CONNECT THE CONCEPT, ABOVE, WITH A CONSENTING NATURAL AREA.*
• **SEVMRATCI** the most interesting felt-sense attractions and senses that you find or that find you[40].(Appendix A: Our Fifty-Four Natural Senses and Sensitivities) • Add this process to an experience you have had in your special area of interest: Art, Creative Writing, Music, Yoga, Parenting, Recovery, Addiction, Renewal, etc.
• **Supportive Reading and Activities:** in a natural area read the next chapter in *Reconnecting with Nature* (or approved course book) including doing its activities and journaling them.
• **Journal Response Form:** For personal growth and reference to describe the value of this section, please complete the Appendix B: Revolutionary Wisdom Response Form: Journal Response Form[41].

THE WARRANTIED FACT CHECK: Visit the warrantied fact list at http://www.ecopsych.com/54warrantfact.docx and insert dates on new facts that you learned or know to this point in the book. Place a check mark on facts you previously dated that you feel have been reinforced or extended.

#25 Expedition Challenge Activity

A human infant, our inner child, is the Planet personified; it embodies the Planet's attractions, life processes, and materials. Its natural tension-relaxation attractions exist to acquaint and connect the infant with the natural world's life-supporting elements and pulse.

With consent, find examples of the 54 natural attraction senses in a natural area and see if you can determine how they supported your life as an infant and now.

For example:

- The infant senses the tension of suffocation to attract the Planet's air. Breathing relaxes this tension and satisfies the Planet's tension for carbon dioxide.
- The infant senses the attraction called thirst because Earth has rain, lakes, and rivers to give it water.
- Thirst attracts the Planet's water and, in turn, the infant's urine relaxes the living planet's tensions for liquid, nitrogen, and support.
- The infant senses hunger because the global life community wants it to eat and live. The infant senses the attractive tension to excrete because Mother Earth needs its by-products as food for other organisms.
- The infant desires mobility to move toward attractive environments that support it and need it.
- The infant feels loneliness because it craves the life-supportive niche and relationships that Earth and relationships.
- The infant senses temperature because it seeks environments that best support its life and vice-versa.
- The infant experiences tensions of sexual desire and nurturing as part of nature's love for it and its kind.
- The infant senses music, form, and color so that it may react to those aspects of Earth. The infant satisfies the living planet's desire to have supportive life relationships.
- The infant feels tensions and senses because, in concert, senses catalyze global and personal balance.
- The infant trusts the now moment because only in it does the whole of life resonate.
- The infant loves life because love is the nature of NNIAAL.

Considerations

Our infant self remains alive in us throughout our lives as our inner nature's biology. It is entirely sentient. Our mind is mainly designed to pay attention to attractions in the natural world for they are its origin and sustenance. The infant within us craves them, for it is them and of them.

Food for thought

To be fully rational, our rationality must honor our natural senses and act accordingly. When it doesn't, our inner nature emotionally us senses abandonment for it is being disconnected from

NNIAAL, the sources of its life. Our comfort-discomfort feelings at any given moment entirely depend upon wisdom about what combinations of logic, symbol-images, attractions, and tensions touch our inner nature.

EXPLORE NOW: VALIDATE. *CONNECT THE CONCEPT, ABOVE, WITH A CONSENTING NATURAL AREA.*
• **SEVMRATCI** the most interesting felt-sense <u>attractions and senses that you find or that find you</u>[40].(Appendix A: Our Fifty-Four Natural Senses and Sensitivities) • Add this process to an experience you have had in your special area of interest: Art, Creative Writing, Music, Yoga, Parenting, Recovery, Addiction, Renewal, etc.
• **Supportive Reading and Activities:** in a natural area read the next chapter in *Reconnecting with Nature* (or approved course book) including doing its activities and journaling them.
• **Journal Response Form:** For personal growth and reference to describe the value of this section, please complete the Appendix B: Revolutionary Wisdom Response Form: Journal Response Form[41].

#26 Expedition Challenge Activity

Let SUNEH-NNIAAL help you answer a question or strengthen yourself.

1) Identify a question you want assistance with.

2) Find a natural attraction in a natural area, an attraction that you can touch (tree, brook, flower, place, air, etc.) and gain its consent to do this activity with it.

3) Get to know this tree as yourself by identifying as many of the 54-senses that you share with it as you can in this moment of the time-space Unified Field.

> -Validate that now is when the life of its and your essence are identical.
>
> -Validate, too, that during the growth of Earth's life through the eons this natural attraction preceded humanity and part of it loved itself into being to your human life in real time.
>
> -Validate that what is attractive to you about this attraction is that you are doing the attracting.

4) Now, while unified with the attraction, continue to put into language what attractions you a
experiencing that could be helpful to your question.

5) In your imagination become the attraction talking to you and telling you the helpful experiences you found in 4).

6) Identify the insights or values you obtained from 1-5 and apply them when appropriate.

EXPLORE NOW: VALIDATE. *CONNECT THE CONCEPT, ABOVE, WITH A CONSENTING NATURAL AREA.*
• **SEVMRATCI** the most interesting felt-sense attractions and senses that you find or that find you[40].(Appendix A: Our Fifty-Four Natural Senses and Sensitivities) • Add this process to an experience you have had in your special area of interest: Art, Creative Writing, Music, Yoga, Parenting, Recovery, Addiction, Renewal, etc.
• **Supportive Reading and Activities:** in a natural area read the next chapter in *Reconnecting with Nature* (or approved course book) including doing its activities and journaling them.
• **Journal Response Form:** For personal growth and reference to describe the value of this section, please complete the Appendix B: Revolutionary Wisdom Response Form: Journal Response Form[41].

Where is your life with regard to the information, above? How does it affect your past and present, your dreams and future?

PART SEVEN

Validating the Powers of Nature's Unified Attraction Field

"The reconnecting with nature process helped me relieve physical and mental pain. I observed it positively affect people suffering from chronic physical or emotional illness."

"It enhanced self-esteem and reduced my dependencies."

"Empowered me to help others help the environment."

"It significantly decreased my stress and depression."

"Powerfully supports me emotionally, spiritually, mentally, and physically."

"Helped me restore my sense of belonging and place."

"Is a form of prayer or 'conscious contact' with my Higher Power."

"Constantly offers life experiences I can trust, I sleep better."

"Instilled a lasting sense of belonging to something worthwhile."

"Nature's purity and beauty rejuvenated my true self."

"Retrained my 'stinkin' thinkin', so annoyment became enjoyment."

"Attained greater resilience with respect to pain, community, learning, self-esteem."

"Became more environmentally and socially active, apathy became positive energy."

SELF-EVALUATION

In the story of Contemporary Society's science and technology relationships, the life of Planet Earth and its balanced wisdom established itself 3-5 billion years before humanity appeared as part of it. The science also discloses that, on average, today, **over 99 percent of our time is spent disconnected from the aliveness of Earth's non-story ways** as we have embedded ourselves in our established, indoor, central thinking, acts, technologies, spiritualities, and institutions. To our loss, this disconnection makes us produce the destructive competition, excessiveness, materialism, injustices, corruption, assaults and disorder effects that we suffer; they are seldom found in Nature. We are blocked from changing this because we learn to deny that **we have become 54-sense emotionally bonded** to the stories that cause these miseries and we can't shift ourselves or them until we transform these ties into organic, nature-connected, relationships. **Good evidence shows that information alone does not accomplish this,** that the GreenWave-54 methods and materials of Organic Psychology empower us to make this transformation when we create time and space for it to do so.

This Part Seven of *Revolutionary Wisdom* is a collection of statements that Parts 1-6 along with their activities and links have presented.

As above, the reader here considers each of the statements and responds to it by identifying what they consider their percent of agreement with it and its significance to them. Note questions, experiences, and examples that you have about these statements.

The Appendixes contain links to information where you can improve percentages that you would like to increase, and you are also welcome to discuss them with a mentor or our staff at 360-378-6313.

PART I: STATEMENTS

1. Objective facts are false when they are not whole life true.

_____% is my percentage of agreement with this statement.
Its significance to me is ( ) strong ( ) moderate ( ) weak

2. Things go better with Nature because it's love of its/our life is an essence of happiness

_____% is my percentage of agreement with this statement.
Its significance to me is ( ) strong ( ) moderate ( ) weak.

3. The unreasonable application of science defines stupid.

_____% is my percentage of agreement with this statement.
Its significance to me is ( ) strong ( ) moderate ( ) weak.

4. Attraction is the essence of love.

_____% is my percentage of agreement with this statement.
Its significance to me is ( ) strong ( ) moderate ( ) weak.

5. All things are connected or attached by attraction.

_____% is my percentage of agreement with this statement.
Its significance to me is ( ) strong ( ) moderate ( ) weak.

6. What we find is attractive in a natural area is what is doing the finding.

_____% is my percentage of agreement with this statement.
Its significance to me is ( ) strong ( ) moderate ( ) weak.

7. The core of Nature's wisdom is that attraction is conscious of what it is attracted to relate to. It is attracted to become more attractive and stronger by producing the diversity of additional attraction relationships that are conscious of what they are attracted to.

_____% is my percentage of agreement with this statement.
Its significance to me is ( ) strong ( ) moderate ( ) weak.

8. The art and science of Organic Psychology empower anyone to know the web-of-life as themselves and vice-versa.

_____% is my percentage of agreement with this statement.
Its significance to me is ( ) strong ( ) moderate ( ) weak.

9. Mysticism wielding the sword of technology spells disaster

_____% is my percentage of agreement with this statement.
Its significance to me is ( ) strong ( ) moderate ( ) weak.

10. We learn to deny that we must constantly act and grow from the energies of our 54 natural senses while they are in contact with the most attractive natural area or thing available.

______% is my percentage of agreement with this statement.
Its significance to me is ( ) strong ( ) moderate ( ) weak.

11. The integrity of the life of the planet we love and live in is at risk[5].

______% is my percentage of agreement with this statement.
Its significance to me is ( ) strong ( ) moderate ( ) weak.

12. Our human life is also the life of the Web of Life's plant, animal, mineral, and energy kingdoms.

______% is my percentage of agreement with this statement.
Its significance to me is ( ) strong ( ) moderate ( ) weak.

13. In the Web of Life, the life of humanity is unique. It can narrate, think, verbalize, question and build relationships through abstract stories while the rest of the web does not have this ability.

______% is my percentage of agreement with this statement.
Its significance to me is ( ) strong ( ) moderate ( ) weak.

14. The life of Earth's and our life's attraction to survive is not imaginary, insensitive or stupid. Rather, it registers that it wants to live.

______% is my percentage of agreement with this statement.
Its significance to me is ( ) strong ( ) moderate ( ) weak.

15. The world is falling apart because our intelligence is indoctrinated to invent, govern and attach the life of our sense of reason to scientifically unreasonable, nature-disconnected stories. These attachments make us desensitized, nature-conquering robots. They motivate us to excessively exploit and conquer the balanced integrity of the life of Nature and Earth's Unified Field, around, in and as us.

______% is my percentage of agreement with this statement.
Its significance to me is ( ) strong ( ) moderate ( ) weak.

16. Because our nature-estranged stories mislead us, resource depletion, species extinction, mental illness, excessive stress, addictions and many other disorders have <u>increased almost 50 percent since Earth Day, 1970</u>[5] and this rate continues to rise.

_____% is my percentage of agreement with this statement.
Its significance to me is ( ) strong ( ) moderate ( ) weak.

17. Moment after moment, we suffer from the separation created by our story world's substitutes for our eons-old, direct, 54-sense contact way of loving, living and learning while in conscious contact with the life of the natural world in a natural area.

_____% is my percentage of agreement with this statement.
Its significance to me is ( ) strong ( ) moderate ( ) weak.

18. Our scientifically inaccurate stories disengage our mentality and interactions from the balanced and beautiful way that the wisdom of Nature and Earth manifests itself as us.

_____% is my percentage of agreement with this statement.
Its significance to me is ( ) strong ( ) moderate ( ) weak.

19. Today, we need another Planet Earth almost half the size of ours to connect with our planet, so it may return to its attraction balanced wellness. Nobody knows where this extra planet can be found or how to connect it with Earth.

_____% is my percentage of agreement with this statement.
Its significance to me is ( ) strong ( ) moderate ( ) weak.

20. Our massive separation from Nature by our stories is the source of our runaway disorders and corruption. This can easily be reversed. All we need is to have enough natural intelligence left so that **we want to** reverse this destructive separation. Then we simply learn how to use the whole life art and science of Organic Psychology to Grok into our consciousness our inherent 54-sense survival wisdom and happily reconnect it to the beautiful life of Earth's self-correcting integrity in a natural area. That enables the natural world's wisdom to help us do what it does best, to bring the life of Earth/us into equilibrium reasonably.

______% is my percentage of agreement with this statement.
Its significance to me is ( ) strong ( ) moderate ( ) weak.

21. It is crucial that we reconnect our nature-divorced thoughts, feelings, relationships, and stories to the life of our planet. This is because the life of Earth is cooperatively unified. It consists of the power of Nature's Unified Field to peacefully organize, correct and balance itself, including us.

______% is my percentage of agreement with this statement.
Its significance to me is ( ) strong ( ) moderate ( ) weak.

22. Webstrings injured in childhood are vulnerable to other individuals who, due to similar childhood injuries, are less sensitive to this vulnerability, and vice versa. This triggers distress (button pushing) between these individuals that Webstring attractions in natural areas can reduce as they replace these vulnerabilities with in-common attraction love as well as promote unity.

______% is my percentage of agreement with this statement.
Its significance to me is ( ) strong ( ) moderate ( ) weak.

23. The joy and benefits we obtain from wondrous childhood or adult experiences in natural areas demonstrate that genuine sensory contact with authentic Nature helps our stories support, rather than compromise, Earth's integrity. This helps us strengthen Earth's organic attraction to regenerate its optimums of life, cooperation, and diversity whose unconditional unification does not produce garbage or rejection, everything belongs.
    ______% is my percentage of agreement with this statement.
    Its significance to me is ( ) strong ( ) moderate ( ) weak.

24. Albert Einstein's GreenWave Unified Field Process (GreenWave-54) is an Organic Psychology tool. It enables our 54-senses to beneficially *Grok* the essence of Nature's life as Planet Earth and vice versa.

______% is my percentage of agreement with this statement.
Its significance to me is ( ) strong ( ) moderate ( ) weak.

25. Grok means to understand and empathize to the extent that something becomes part of our sense of self-knowing that we exist in its embodiment as well as that we can sense ourselves as it being us. This is similar to "being at one with nature," how things in the

web of life know each other through attraction relationships rather than by stories about them.

_____% is my percentage of agreement with this statement.
Its significance to me is ( ) strong ( ) moderate ( ) weak.

26. The purpose of this book's expedition adventures is it to teach you to know *how* and *why* to use the Organic Psychology GreenWave-54 tool to help us address our problems. Without doing both, we increasingly travel our wayward path.

_____% is my percentage of agreement with this statement.
Its significance to me is ( ) strong ( ) moderate ( ) weak.

This activity helps increase a person's GreenWave-54 effectiveness

1) Find in a natural space where you sense at least one attractive example of the paragraph's message taking place. Then
2) Grok the natural attraction using one or more of your favorite GreenWave-54 nature-connection activities that you have learned.

For example

A. What attractive senses and feelings do I experience in the life of this area?
http://www.ecopsych.com/MAKESENSEWALK.docx

B. Ask the attraction, "Who are you without your label/name?" After it responds ask it, or ask you, "Who am I without my label/name?"

C. With the attraction you found, do the activity at www.ecopsych.com/giftearthday1.html

D. Ask yourself "What would I sense and feel if my ability to register this attraction connection was taken away from me?

The activity, above, helps the life of Earth prevent you from misleading yourself and others by only giving lip service to what you think you know and then depending upon that same half-true source of information in an industrial society that is increasingly producing Earth Misery.

_____% is my percentage of agreement with this statement.
Its significance to me is ( ) strong ( ) moderate ( ) weak.

The positive outcome that you experience from Grokking is the self-evident GreenWave-54 fact that empowers you appropriately guide your life experiences and helps others do the same.

_____% is my percentage of agreement with this statement.
Its significance to me is ( ) strong ( ) moderate ( ) weak.

"Choose a job you love, and you will never have to work a day in your life."- Confucius

PART II: REASONABLENESS

Use this evidence-based reality check. At this moment:

1. You are alive. ( ) yes ( ) no

_____% is my percentage of agreement with the statement above.
Its significance to me is ( ) strong ( ) moderate ( ) weak.

2. You can understand these words as they appear via this technology be it a screen, book or sound system. ( ) yes ( ) no

_____% is my percentage of agreement with the statement above.
Its significance to me is ( ) strong ( ) moderate ( ) weak.

3. This technology results from folks reasonably researching for evidence that disclosed if its components were dependable and its results repeatable. ( ) yes ( ) no

_____% is my percentage of agreement with the statement above.
Its significance to me is ( ) strong ( ) moderate ( ) weak.

4. You are presently geographically located in, not on, Planet Earth, enveloped by the life of its biosphere and embedded under its miles of sunshine, air, clouds and stratosphere above and around you. ( ) yes ( ) no

_____% is my percentage of agreement with the statement above.
Its significance to me is ( ) strong ( ) moderate ( ) weak.

5. Scientifically, the time and life of our Universe, also known as Nature or the Cosmos, was born in the Big Bang. ( ) yes ( ) no

_____% is my percentage of agreement with the statement above.
Its significance to me is ( ) strong ( ) moderate ( ) weak.

6. The life of Earth and your life is located at the same time, space and ways of the Universe because they are identical. The main exception is that your life can create, narrate, understand and act-out stories while Earth is mute and can not do this. ( ) yes ( ) no

_____% is my percentage of agreement with the statement above.
Its significance to me is ( ) strong ( ) moderate ( ) weak.

7. You can sense and feel that:
   a. You are *conscious.*
   b. Your sense of reason *senses that it can be reasonable.*
   c. You reasonably *sense that you want to breathe.*
   d. You reasonably *sense that your self* wants to be alive at the next moment.
   e. You reasonably *sense that you want to trust people, places, and things*
   f. You reasonably *sense that you want to be attractive and loved.*

( ) yes ( ) no

_____% is my percentage of agreement with the statement above.
Its significance to me is ( ) strong ( ) moderate ( ) weak.

8. A story you have may have learned says that the six senses, above, are not included in our alleged only five senses (touch, taste, smell, sight, and sound). ( ) yes ( ) no

_____% is my percentage of agreement with the statement above.
Its significance to me is ( ) strong ( ) moderate ( ) weak.

9. You are reasonable enough to recognize, from the above, that the story that says we know the world through only five senses is not accurate. ( ) yes ( ) no

_____% is my percentage of agreement with the statement above.
Its significance to me is ( ) strong ( ) moderate ( ) weak.

10. You are reasonable enough to recognize that your reading experience, above, provided you with self-evident information that was more accurate than the story that says that we know and relate to the world through five senses. ( ) yes ( ) no

_____% is my percentage of agreement with the statement above.
Its significance to me is ( ) strong ( ) moderate ( ) weak.

11. You are reasonable enough to recognize that although Nature communicates it does not accomplish this through the spoken or written word-verbalization, narration or literacy of our stories. ( ) yes ( ) no

_____% is my percentage of agreement with the statement above.
Its significance to me is ( ) strong ( ) moderate ( ) weak.

12. The life of Earth and its eons is story-less. You have not, and no scientific evidence has discovered, members of the plant, animal, mineral or energy kingdoms, now or ever, narrating, vocalizing or writing words, sentences, paragraphs or books. ( ) yes ( ) no

_____% is my percentage of agreement with the statement above.
Its significance to me is ( ) strong ( ) moderate ( ) weak.

13. You are reasonable enough to recognize that Earth changed with the geologically recent appearance of humanity. Humans were and are gifted with the ability to communicate and build relationships using stories. Stories help us shortcut and mediate the way our senses first-hand experience how the world works. Stories abstract that experience. For example, the story about how you can quench your thirst by drinking water is different from being motivated to drink water. The story alone does not realistically satisfy your body's need for water nor your thirst while the act does satisfy both without producing garbage. ( ) yes ( ) no

_____% is my percentage of agreement with the statement above.
Its significance to me is ( ) strong ( ) moderate ( ) weak.

14. We have created stories that socialize us to build, attach to and live mostly in scientifically powerful, nature-separated and conquering, indoor societies. This emotionally binds or addicts our senses to nature-isolated methods and materials while it also disconnects us from consistent sensory contact with the wisdom of authentic Nature. ( ) yes ( ) no

_____% is my percentage of agreement with the statement above.
Its significance to me is ( ) strong ( ) moderate ( ) weak.

15. Authentic Nature is the fountainhead of authority in how its whole-life, self-organizing and self-correcting ways work in balance, purity, and beauty around and within us

without producing garbage. In the life of Nature, everything belongs. ( ) yes ( ) no

_____% is my percentage of agreement with the statement above.
Its significance to me is ( ) strong ( ) moderate ( ) weak.

16. The destructive effects of humanity for centuries excessively acting out our nature-disconnected stories in unbalanced ways are now obvious and measurable[5]. ( ) yes ( ) no

_____% is my percentage of agreement with the statement above.
Its significance to me is ( ) strong ( ) moderate ( ) weak.

11. GreenWave-54 works sensibly by helping us be open minded to hidden yet self-evident facts, information and connections that our 54 natural senses gather as they learn how to Grok the life of nature in a natural area. This process extends our limited 5-sense sensibility to make, on average, 85 percent greater sense of our relationships with our planet, each other and our body, mind, and spirit. ( ) yes ( ) no

_____% is my percentage of agreement with the statement above.
Its significance to me is ( ) strong ( ) moderate ( ) weak.

12. The art and science of the Unified Field process (GreenWave-54) produce trustable facts because its evidence-based thinking and relationships exclude mystical, misleading or unsubstantiated stories. Its empirical source is nature, not fantasy. ( ) yes ( ) no

_____% is my percentage of agreement with the statement above.
Its significance to me is ( ) strong ( ) moderate ( ) weak.

***"The whole of science is nothing more than a refinement of everyday thinking."* - Albert Einstein**

PART III: ORGANIC PSYCHOLOGY RATIONALE

In continual review use your inherent abilities to identify and Grok the real-life source and essence of Nature, in and around you and others in a natural area.

They help you unconditionally experience:

1. The life of Nature's wordless wisdom beautifully speaks for itself.

_____% is my percentage of agreement with the statement above.
Its significance to me is ( ) strong ( ) moderate ( ) weak.

They help you unconditionally experience:

2. The silence of the life of Nature takes you to the core of love.

_____% is my percentage of agreement with the statement above.
Its significance to me is ( ) strong ( ) moderate ( ) weak.

They help you unconditionally experience:

3. The attractiveness of natural attraction being conscious of what it is attracted to.

_____% is my percentage of agreement with the statement above.
Its significance to me is ( ) strong ( ) moderate ( ) weak.

They help you unconditionally experience:

4. The greatest truth in your life is the sensation that you register right now.

_____% is my percentage of agreement with the statement above.
Its significance to me is ( ) strong ( ) moderate ( ) weak.

They help you unconditionally experience:

5. We are each naturally born to love and be loved.

_____% is my percentage of agreement with the statement above.
Its significance to me is ( ) strong ( ) moderate ( ) weak.

They help you unconditionally experience:

6. The life of Nature is the intelligence that unconditionally loved you into being.

_____% is my percentage of agreement with the statement above.
Its significance to me is ( ) strong ( ) moderate ( ) weak.

They help you unconditionally experience:

7. We live within, not just on, our living planet as its sanity flows through us.

_____% is my percentage of agreement with the statement above.
Its significance to me is ( ) strong ( ) moderate ( ) weak.

They help you unconditionally experience:

8. As it has since the beginning of the time and space that it makes, the natural world's exquisite genius purely organizes, corrects and balances itself.

_____% is my percentage of agreement with the statement above.
Its significance to me is ( ) strong ( ) moderate ( ) weak.

They help you unconditionally experience:

9. The closest wilderness to you at any time lies within you. This means that you, along with everything else, are attached to all that came before and will remain attached to all that follows.

_____% is my percentage of agreement with the statement above.
Its significance to me is ( ) strong ( ) moderate ( ) weak.

They help you unconditionally experience:

10. All of the life of Nature and Earth lives and transforms into wordless forms of the life cycle that we learn to call death.

_____% is my percentage of agreement with the statement above.
Its significance to me is ( ) strong ( ) moderate ( ) weak.

They help you unconditionally experience:

11. Nature creates its pure and harmonic optimums of life, diversity, and cooperation without producing garbage, excessiveness or abuse.

_____% is my percentage of agreement with the statement above.
Its significance to me is ( ) strong ( ) moderate ( ) weak.

They help you unconditionally experience:

12. The personal and interpersonal love and support from the life of Earth that you share and naturally deserve.

_____% is my percentage of agreement with the statement above.
Its significance to me is ( ) strong ( ) moderate ( ) weak.

They help you unconditionally experience:

13. The facts of organic psychology art and science bring to mind and validate what your heart already knows.

_____% is my percentage of agreement with the statement above.
Its significance to me is ( ) strong ( ) moderate ( ) weak.

PART IV: THIS REVOLUTIONARY WISDOM EXPEDITION BOOK

1. Recognizes that our lives depend on our advanced objective science and technology that for accuracy must omit the mystical. However, for total accuracy, it must also include undeniable self-evidence that our 54 natural senses register in natural areas as facts of life.

_____% is my percentage of agreement with the statement above.
Its significance to me is ( ) strong ( ) moderate ( ) weak.

2. Revolutionary Wisdom includes self-evident facts from research during eighty years of intimate relationships with the life of nature in and around expedition groups, indoors and in natural areas.

_____% is my percentage of agreement with the statement above.
Its significance to me is ( ) strong ( ) moderate ( ) weak.

3. Embraces undeniable facts that are obvious to you because you experience them, such as you are reading these words right now and that you are attracted to breathing.

_____% is my percentage of agreement with the statement above.
Its significance to me is ( ) strong ( ) moderate ( ) weak.

4. Produces the beneficial results of year-long, expedition education community explorations of the life of Earth in and around us, a multisensory process Mike Cohen developed from 1965-1985 in the Expeditions' objective search for Mother Earth.

_____% is my percentage of agreement with the statement above.
Its significance to me is ( ) strong ( ) moderate ( ) weak.

5. Revolutionary Wisdom occurs within the lively dance of our scientifically substantiated Universe. Its 4 trillion degree, Big Bang birth some 13.8 billion years ago included a unifying attraction field that ever since attracts everything to everything else including the past to the future.

_____% is my percentage of agreement with the statement above.
Its significance to me is ( ) strong ( ) moderate ( ) weak.

6. The Universe has continually cooled, diversified and created itself into its next moment of time and space relationships right to this moment. Its dance diversifies into the world that, as part of it, our 54-senses experience and register.

_____% is my percentage of agreement with the statement above.
Its significance to me is ( ) strong ( ) moderate ( ) weak.

7. Attraction unifies the dance of Nature/Earth's eons, in every moment. Like the mathematical sequence of 0-9, each and everything is attached to what came before it and to what comes after it. Objective evidence collected during 85 years of research since 1925 has substantiated this universal fact of life theory. In 2012 that same painstaking scientific process affirmed the elusive Higgs Boson origin of every moment's Unified Field attraction energies and the time and space they produce along with gravitational and electromagnetic and other energy fields.

_____% is my percentage of agreement with the statement above.
Its significance to me is ( ) strong ( ) moderate ( ) weak.

8. The dance of the global life system communicates with us and vice versa for we are part of it and to be part of a system we have to be in communication with it. We GreenWave instinctively experience that communication as survival, as our attraction for the system and us to live. Our whole life intelligence knows that if the life of Earth sickens or dies, so do we.

_____% is my percentage of agreement with the statement above.
Its significance to me is ( ) strong ( ) moderate ( ) weak.

PART V: WHO SAYS WHAT?

1. Concerning the example of the runaway car driving into a family picnic in a natural area, we are taught to ignore that the appropriate brake and steering system that we need **is alive and well in the natural area that the car is destroying.**

_____% is my percentage of agreement with the statement above.
Its significance to me is ( ) strong ( ) moderate ( ) weak.

2. We learn to disavow that the area's GreenWave-54 brakes and guidance work because **they simultaneously operate in us as well as around us in natural areas.**

_____% is my percentage of agreement with the statement above.
Its significance to me is ( ) strong ( ) moderate ( ) weak.

3. We are told to reject that **all things are attracted to contribute to GreenWave-54 remedies** as well as attracted to be guided by them.

_____% is my percentage of agreement with the statement above.
Its significance to me is ( ) strong ( ) moderate ( ) weak.

4. **The folks who are teaching us to omit or disavow GreenWave-54 are the same folks who 1) don't know how to drive the car safely, 2) are profiting by selling us the car and 3) are daily increasing our earth misery catastrophe**[5]. Unless we have been brainwashed, how can we possibly trust or follow their information if we already know the adverse results?

_____% is my percentage of agreement with the statement above.
Its significance to me is ( ) strong ( ) moderate ( ) weak.

5. These folks promise things like "more economic growth" while Earth misery Day[5] shows that the **life of Earth is already bankrupt, that our deficit and misery increases year after year** and that the Web-of-Life, including us, is increasingly the victim of this excessive spending.

_____% is my percentage of agreement with the statement above.
Its significance to me is ( ) strong ( ) moderate ( ) weak.

6. Choosing to add GreenWave-54 to our lives feels and heal better. It helps us protect ourselves and each other from injury while we eliminate the core of the problem.

_____% is my percentage of agreement with the statement above.
Its significance to me is ( ) strong ( ) moderate ( ) weak.

7. **Some folks recognize that GreenWave-54 is prayer, hope and stress management in action.** Some honor the integrity of their GreenWave attractions by labeling themselves nature lover, agnostic, environmentalist, atheist, pantheist, humanist, eco-artist, scientist, etc. However, because GreenWave-54 is <u>illegally withheld,</u> very few of them are aware that the art and science of GreenWave-54 education, counseling, and healing with nature is a key supportive tool and unifying that they best add to every aspect of their life.

_____% is my percentage of agreement with the statement above.
Its significance to me is ( ) strong ( ) moderate ( ) weak.

8. **Most scientists agree that our Industrial way of life is runaway.** It can't stop increasing climate change, social injustice, relationship disorders, species extinction and mental illness without producing additional problems. Without using GreenWave-54, these same scientists don't know how or why they and we continue to do this or how to cease. We emotionally and physically suffer today because this shortcoming is typical of our leaders, interactions, families and most of us, locally and globally.

_____% is my percentage of agreement with the statement above.
Its significance to me is ( ) strong ( ) moderate ( ) weak.

9. **Today our choice is clear,** Either we:

   1) Continue omitting GreenWave-54 organic psychology from our well-intentioned training and relationships and go on producing today's deadly runaway and immoral results or

   2) We reverse this catastrophe by adding GreenWave-54 to everything we do and let our strengthened and supported GreenWave wisdom help us enjoy and increase the delightful unifying sanity of personal, social and environmental well-being.

   3) We abandon Earth and establish artificial human communities in outer space

Statement number you agree with: ____

_____% is my percentage of agreement with the "choice" statement above.
Its significance to me is ( ) strong ( ) moderate ( ) weak.

10. **GreenWave-54 works** because it puts to work the unifying and self-correcting powers of nature's attraction essence that exists in our daily lives every moment. It includes electricity, magnetism, gravity and sunlight attraction along with our 54-senses[27] that can register and monitor their effects.

_____% is my percentage of agreement with the statement above.
Its significance to me is ( ) strong ( ) moderate ( ) weak.

PART VI: WHOLE LIFE QUESTIONS

NOTE: In the section below, for each question, see if you can find an example of its answer from:

a) Your experiences,
b) Parts 1-3 and their links,
c) Or Simply answer the question.

For each statement below add:

- ✓ Your percentage of knowing,
- ✓ And the significance of the statement to you. (Examples are provided for numbers 1 and 2.)

1. What is the most accurate and trustable form of information? Why?

_____% **My percent of knowing the answer to this question. Give an example of it.**
Its significance to me is ( ) strong ( ) moderate ( ) weak.

2. What is the greatest truth in your life that you can trust? (Clue: the answer is not God, love, honesty or nature.)

_____% **My percent of knowing the answer to this question. Give an example of it.**
Its significance to me is ( ) strong ( ) moderate ( ) weak.

3. What is the key factor that makes humanity different or separate from nature?
4. How and why doesn't nature produce any garbage or toxic waste?

5. How can you/we be sure that Planet Earth is our Other Body?
6. What is the point source of contemporary society's environmentally and socially destructive ways?
7. How do we know if Nature and Earth are intelligent?
8. What are the difference between a fact, thought, and a feeling?
9. What produces the wanting void in our psyche, the discomfort, greed, and loneliness that fuels most human and environmental disorders?
10. Why do we deny that we are addicted to stories, technologies, and relationships that separate us from nature's balanced ways and restorative powers?
11. If we learn to be who we are, what factor in modern education teaches us to produce today's unsolvable problems?
12. To be part of a system, you have to be in communication with it in some way. We are part of the global life system and vice versa; how does it communicate with us and we with it?
13. What is the essence that makes nature's perfection work?
14. What is attractive about fear, stress or pain?
15. Where does death exist in nature?
16. Does nature have a direction or purpose?
17. What are five steps to letting nature help you reduce destructive attachments
18. What value is there in safely feeling closer nature?
19. Do we deserve to have good feelings? Why?
20. Why do we continue to assault nature and people when it doesn't make sense, and we don't like doing it or its effects?
21. Can one be sane if they are a good citizen of an insane society?
22. What is consciousness and who invented it?
23. Do miracles happen in contradiction to nature?
24. What is the relationship between nature and the human spirit?
25. Since nature produces no garbage, is nature unconditional love?
26. How and where do you collect self-evidence?
27. Can you cite a model, community or process that successfully produces sustainability?
28. How do our senses know that Planet Earth a living organism? Is it?
29. How much of our ability to sense and feel do we inherit from nature?
30. What is the relationship between our increasing personal social and environmental problems?
31. What makes Natural Attraction Ecology be a pure and whole, rather than limited, objective science?
32. Where do we physically live in the solar system?
33. What is time in nature
34. How do we know if Nature is alive?
35. What is natural homeostasis in nature and how does it work?
36. Why do we excessively conquer or exploit nature?

37. Why is it best to identify nature as a dance?
38. Does duality exist in nature?
39. What is the difference between nature and life?
40. What is NNIAAL
41. Are the Higgs Boson and NNIAAL identical?
42. What is an Earth Avatar?
43. What is the difference between nature and the Standard Universe?
44. In our Other Body natural attraction Dance when is nature repulsive?
45. In the macro Natural Attraction Dance of the Unified Field, what are the names given to its micro pulsating resonances and fluctuations?
46. What is the purpose and contribution of "Validating?"
47. What is the significance of "Tropicmaking."
48. Why are webstrings often subconscious and how do we bring them into awareness?
49. How many natural senses do we have? Can we think with them? How?
50. How many natural senses can you name that you can know and learn from?
51. How does nature within you know how to relate responsibly to nature in others and the environment?
52. If life has a purpose, what is it?
53. Is our innate ability to sense and feel of, by and from nature?
54. Do our formal education or leaders competently address the above questions?
55. Why don't our cultural stories and dollar bills produce balance and purity, like nature works?
56. Do our unsolvable problems result from the difference between how we think and how nature works?
57. In nature, why don't two plus two equal four?
58. Where in nature do you find life abstracted?
59. What in Nature is not conscious on some level?
60. What parts of Nature is prejudiced against Nature?
61. What is the relationship between natural attractions and consciousness?
62. Is nature a form of perfection people can achieve and if so, how.
63. What is the element that produces environmentally and socially destructive economic relationships?
64. Since we are part of Nature, what is the significant difference that makes us destroy the environment while everything else in Nature usually strengthens it?
65. How can we restore to our thinking the missing 49 sensory intelligences that contemporary society has buried in our subconscious?
66. Why does modern society often identify a person's love of nature as "escapist recreation and fuzzy thinking" rather than "peaceful, reasonable re-creation?"
67. What are four additional words that mean the same thing as "natural attraction"?
68. What is the essential contribution of the Unified Field Equation?
69. What is the value of identifying and relating through singularities?

70. What unique does Scientific methodology, the Big Bang, math logic and the Unified Field have in common?
71. What are the benefits of recognizing that God is love and is found everywhere?
72. What is the contribution of the GreenWave addition to the Web-of-Life model?
73. Are the effects of applying scientific facts a valid way of measuring them?
74. How does scientific logic show that the Universe is alive?
75. What sense makes sense to the other 53 senses by Validating?
76. Why does Scientific Methodology presently omit the contribution of sensory subjectivity?
77. What is the value of attraction at the beginning of time, or before, being conscious of its attraction to support life?
78. How does a yellow zipper relate to a blue and orange sided piece of paper?
79. What is SUNEH-NNIAAL 54?
80. What is CRL?

Check your answers to these 80 questions in Appendix D: Whole Life Question Responses.

What are questions missing that you think should be asked?

Where is your life with regard to the information, above? How does it affect your past and present, your dreams and future?

A HYPOTHESIS: You can beneficially learn to personify the GreenWave and teach others to do the same to the benefit of all.

Appendix A: Our Fifty-Four Natural Senses and Sensitivities

This list explains how, sense by sense in 54-sense resonance, the GreenWave Natural Attraction Unified Field (GreenWave-54) connects with and unifies itself in us, through us and with people and places around us. By putting these senses into scientific stories and labels the list enables our sense of language (sense #39) to consciously (sense #42) and reasonably (sense #43) translate into and engage in reasonable stories that connect us to the life and love of Nature/Earth's moment-by-moment, self-correcting survival process (sense #54). GreenWave-54 is the outcome of the author's 51 years of living this organic experience in natural area space and time with his trained, evidence-based knowledge and awareness.

Note: Read http://www.ecopsych.com/insight53senses.html and *The Seven Mysteries of Life* by Guy Murchie.

The Radiation Senses

1. Sense of height and sight, including polarized light.
2. Sense of seeing without eyes such as heliotropism or the sun sense of plants.
3. Sense of color.
4. Sense of moods and identities attached to colors.
5. Sense of awareness of one's visibility or invisibility and consequent camouflaging.
6. Sensitivity to radiation other than visible light including radio waves, X rays, etc.
7. Sense of Temperature and temperature change.
8. Sense of season including the ability to insulate, hibernate and winter sleep.
9. Electromagnetic sense and polarity which includes the ability to generate current (as in the nervous system and brain waves) or other energies.

The Feeling Senses

10. Hearing including resonance, vibrations, sonar and ultrasonic frequencies.
11. Awareness of pressure, particularly underground, underwater, and to wind and air.
12. Sensitivity to gravity.
13. The sense of excretion for waste elimination and protection from enemies.
14. Feel, particularly touch on the skin.
15. Sense of weight, gravity, and balance.
16. Space or proximity sense.
17. Coriolus sense or awareness of effects of the rotation of the Earth.
18. Sense of motion. Body movement sensations and sense of mobility.

The Chemical Senses

19. Smell with and beyond the nose.

20. Taste with and beyond the tongue.
21. Appetite or hunger for food, water, and air.
22. Hunting, killing or food obtaining urges.
23. Humidity sense including thirst, evaporation control, and the acumen to find water or evade a flood.
24. Hormonal sense, as to pheromones and other chemical stimuli.

The Mental Senses

(25-27 are attractions that "say" ***seek additional natural attractions*** to support well-being.)

25. **Pain, external and internal.**
26. **Mental or spiritual distress.**
27. **Sense of fear, dread of injury, death or attack**

28. Procreative urges: sex awareness, courting, love, mating, paternity and raising young.
29. Sense of play, sport, humor, pleasure, and laughter.
30. Sense of physical place, navigation senses including detailed awareness of land and seascapes, of the positions of the sun, moon, and stars.
31. Sense of time and rhythm.
32. Sense of electromagnetic fields.
33. Sense of weather changes.
34. Sense of emotional place, of community, belonging, support, trust, and thankfulness.
35. Sense of self including friendship, companionship, and power.
36. Domineering and territorial sense.
37. Colonizing sense including compassion and receptive awareness of one's fellow creatures, sometimes to the degree of being absorbed into a superorganism.
38. Horticultural sense and the ability to cultivate crops, as is done by ants that grow fungus, by fungus who farm algae or birds that leave food to attract their prey.
39. **Language and articulation sense** used to express feelings and convey information in every medium from the bees' dance to human stories and literature.
40. Sense of humility, appreciation, ethics.
41. Senses of form and design.
42. **Sense of Reason,** including memory and the capacity for logic and science.
43. **Sense of mind and Consciousness.**

(*39, 42, 43 are the* ***CRL core of Organic Psychology***)

44. Intuition or subconscious deduction.
45. Aesthetic sense, including creativity and appreciation of beauty, music, literature, form, design, and drama.
46. Psychic capacity such as foreknowledge, clairvoyance, clairaudience, psychokinesis, astral

projection and possibly certain animal instincts and plant sensitivities.
47. Sense of biological and astral time, awareness of past, present and future events.
48. The capacity to hypnotize other creatures.
49. Relaxation and sleep including dreaming, meditation, brainwave awareness.
50. Sense of pupation including cocoon building and metamorphosis.
51. Sense of excessive stress and capitulation.
52. Sense of survival by joining a more established organism.
53. Spiritual sense, including conscience, capacity for sublime love, ecstasy, a sense of sin, profound sorrow and sacrifice.
54. Sense of homeostatic unity, of natural attraction as the singular mother/seed essence of all our other senses *(and everything else that "Tree of Life" singularity was and remains attracted to create, unify and support moment-by-moment as Albert Einstein's Big Bang Unified Field (Higgs Boson) that was verified in 2012 A.D.).*

Appendix B: Revolutionary Wisdom Response Form

Note: if you have previously utilized the 12 Interact Catalysts in a previous course, please add these, whenever possible to the response form below.

The goal of this book/course is to help you use the art of Organic Psychology to resolve problems and create well-being in our technology-attached lives scientifically. It achieves this by holistically applying our civilization's core truth, evidence-based information process discovered by Thales of Miletus, circa 600 B.C. He successfully omitted the mystical and supernatural from accounts of nature. Simply do the same here to achieve this book's goal for yourself in your responses to the questions, below. Challenge yourself to grow. Try to answer questions that you can't answer and ask other course participants for help with them as well as **find them via our Project NatureConnect search engine:** http://www.ecopsych.com/ksearchengine.html

Section Response Form Chapter ________ Section #_______ or other information response

Date Place Weather:

A. What happened in the Groking activities? How and what did you 54-sense, feel and learn[27] from this section (See Appendix A: Our Fifty-Four Natural Senses and Sensitivities)? What senses were most active or attractive to you?
B. What experiences or beliefs of yourself or others does this section affect? How will you modify the terms, labels, and concepts in them to blend them into in this section?
C. How would you feel about having this section's experience taken away from you?
D. What did you discover that helped you recognize contemporary society has misled itself so we "habitually" injure ourselves and others.
E. How would you teach what you learned from this experience to a group of people or dolls?
F. Can you find an example of where this experience is inaccurate with respect to whole life, 54-sense science? Note that, in theory, this is impossible.
G. Write one or two keywords that convey the important things you learned from this section
H. Write one or more complete, single, short, power sentence "quotes" that express an attractive contribution that this experience makes.
I. List two or more quotes you can find on the internet that support this section.
J. Make a flash card with a question or questions on it that this section answers. Put the answers on the reverse side of the card and test yourself with the questions when you think it will be helpful. Your cards will cover most of the questions on the Comprehensive Exams you are required to take for degrees. Add answers to your cards as you progress through

additional sections. These cards will also make this book's facts easy to review.

K. LATER While your story mind sleeps, your 54-sense mind fills new paths of consciousness made by this activity. Note upon awakening whether any changes have occurred immediately, or in the next few days, with respect to the way you think, feel or act.

Appendix C: Resources

Resources *http://www.ecopsych.com*

1. Webstrings *webstrings1000.html*
2. Duality: Four vs. five vs. nine leg thinking *nineleg.html*
3. 54 natural senses *insight53senses.html*
4. Self-evidence is pure truth *journalpeak.html*
5. GO to GG *counseling.html*
6. Web of life model *ksanity.html*
7. Ecology-based on natural attraction *http://www.naturalattractionecology.com*
8. Planet Earth is a living organism *livingplanetearthkey.html*
9. Big Bang *natural attraction seed* attributes *mjcohen22.html*
10. Warranty of key information accuracy journalwarranty.html
11. Uncomfortable natural senses (25-27) are attractions *mjcohen22.html*
12. Disconnection role of Institutions and Socialization *box.html*
13. Nature is non-literate and non-dualistic Dance *journalessence.html*
14. Who or What are you? *thesisquote6.html*
15. Twenty axioms *http://www.naturalattractionecology.com/?page_id=30*
16. Seeking Consent *amental.html*
17. The toxic CRL triad *journalessence.html*
18. NNIAAL *earthstories101.html*
19. New Brain *counseling.html*
20. Old Brain *counseling.html*
21. Higgs Boson significance *journalstfrancis.html*
22. Tropicmaking *index.html* or *ksanity.html*
23. Nature and Dualism *journalessence.html*
24. Labels misrepresent nature *ksanity2.html*
25. Source of and rationale for good feelings *http://www.naturalattractionecology.com*
26. Nature as Higher Power *nhpbook.html*
27. How and why the course activity transformation process works. *transformation.html*
28. Relationship of God to the natural attraction universe "seed" *mjcohen22.html*
29. The source of greed, addiction, abusiveness, depression, disorders *box.html*
30. What are empirical facts? *ksanity.html*
31. Webstring attachments to stories and technologies. *journalessence.html*
32. Prejudice against nature *prejudicebigotry.html*
33. Institutions capture natural senses journalinstitution.html
34. Are the numbers one and zero true? *ksanity2.html*

35. Program History: left-hand diversity, Sunnyside Gardens, Progressive Education *ksanity2.html* *history.html*
36. The Big Bang Universe is alive universealive.html
37. Corruption of the human dancer and the dance opnaeinfo.html
38. Humanity and the flowing river dance of natural attractions *mjcohen22.html*
39. The NNIAAL dance and pulse of natural attraction *earthstories101.html*
40. The Einstein Unified Field Equation *journalaliveness.html*
41. Attraction is Conscious of What it is Attracted To *GREENWAVEBETA.docx*
42. Answers to questions 1-78 *eco800set18examanswers.html*
43. Index of Project NatureConnect key web pages annotatedpages.html

Appendix D: Whole Life Question Responses

Answers to *Whole Life Questions*, from Michael J. Cohen, Ph.D., Founder and President of Project NatureConnect ECHN courses and programs since 1965.

1. What is the most accurate and trustable form of information? Why?

Answer: Personal experience because it registers directly on one or more of our 54 natural senses.

2. What is the greatest truth in your life that you can trust? (Clue: the answer is not God, love, honesty or nature.)

Answer: Information and things we experience/register in the immediate moment for that is the only time our natural senses operate.

3. What is the key factor that makes humanity different or separate from nature?

Answer: Humanity can operate and relate using inaccurate or nature-disconnecting stories.

4. How and why doesn't nature produce any garbage or toxic waste?

Answer: Nature is attraction based. All things in the natural world are attractive and belong moment after moment, so no garbage is produced. There are no organic negatives in Nature

5. How can you/we be sure that Planet Earth is our Other Body?

Answer: Our "story body" can ask our "other body" to disconnect from Earth by not breathing. If the feeling that follows attracts us to re-connect with Earth by breathing again, it signals that our other body is part of Planet Earth.

Go to a natural area and find examples there of the two most attractive of the five answers above.

6. What is the point source of contemporary society's environmentally and socially destructive ways?

Answer: The senses of Reason and Consciousness becoming attached to nature inaccurate or nature-disconnecting stories.

7. How do we know if Nature and Earth are intelligent?

Answer: It is intelligent to find life attractive and support it, and that is what they do.

8. What are the difference between a fact, thought, and a feeling?

Answer: A fact is a story that is reasonable, and when actualized it does what it says it would do. A thought is a story or feeling, true or untrue. A feeling is a sensation that tries to motivate or modify behavior.

9. What produces the wanting void in our psyche, the discomfort, greed, and loneliness that fuels most human and environmental disorders?

Answer: Disconnection from fulfillment in nature.

10. Why do we deny that we are addicted to stories, technologies, and relationships that separate us from nature's balanced ways and restorative powers?

Answer: Disconnecting from our life support system is so stupid that our intelligence denies that it has done this and that prevents us from experiencing the pain of our shame.

Go to a natural area and find examples there of the two most attractive of the five answers above.

11. If we learn to be who we are, what factor in modern education teaches us to produce today's unsolvable problems?

Answer: Being rewarded for and attached/addicted/bonded to thinking and living out nature-disconnecting stories.

12. To be part of a system, you have to be in communication with it in some way. We are part of the global life system and vice versa; how does it communicate with us and we with it?

Answer: Through 54 natural attraction senses and sensations.

13. What is the essence that makes nature's perfection work?

Answer: Natural attraction, Love, Life, Unified Field, Big Bang.

14. What is attractive about fear, stress or pain?

Answer: It attracts us to seek more attractions and support.

15. Where does death exist in nature?

Answer: There is only a transformation and transition recycling of life; death is not attractive

Go to a natural area and find examples there of the two most attractive of the five answers above.

16. Does nature have a direction or purpose?

Answer: To support its life including Earth's web-of-life.

17. What are five steps to letting nature help you reduce destructive attachments

Answer:

-Find your strongest attraction in a natural area;

-Get consent to visit it, then thank it;

-Discover what other attractions or stories come into consciousness;

-Validate them and their good feelings;

-Sleep on the experience.

18. What value is there in safely feeling closer nature?

Answer: Supportive attractions come into play and register.

19. Do we deserve to have good feelings? Why?

Answer: We are personifications of our natural attractions and their origins in the Unified Field/NNIAAL. Good feelings in natural areas rewardingly signal that we are in a mutually beneficial relationship with attractions there.

20. Why do we continue to assault nature and people when it doesn't make sense, and we don't like doing it or its effects?

Answer: We have lost webstring contact with parts of nature and people, so we insensitively trespass them for fulfillment to replace this loss.

Go to a natural area and find examples there of the two most attractive of the five answers above.

21. Can one be sane if they are a good citizen of an insane society?

Answer: Only when they are helping society make webstring contact with nature to become saner.

22. What is consciousness and who invented it?

Answer: One of the 54 natural attraction senses that display experiences and stories on its screen when they have enough energy to register there.

23. Do miracles happen in contradiction to nature?

Answer: No, if you examine them closely they can be seen as many of our 54 webstrings in action.

24. What is the relationship between nature and the human spirit?

Answer: They are identical sensory feelings except that the human spirit can also register as a story.

25. Since nature produces no garbage, is nature unconditional love?

Answer: Yes, on a macro level so that it includes all of the web-of-life.

> **Go to a natural area and find examples there of the two most attractive of the five answers above.**

26. How and where do you collect self-evidence?

Answer: By experiencing and validating it as it registers on your webstring senses.

27. Can you cite a model, community or process that successfully produces sustainability?

Answer: Any setting in nature that included humanity and has not been contaminated.

28. How do our senses know that Planet Earth a living organism? Is it?

Answer:

> -In any given moment the essence of all things in the Universe are identical as natural attraction manifesting itself as the Universe. For this reason, if any one thing is alive in a moment, all things are also alive. Any moment that a person knows they are alive means

Earth and the Universe are alive including the Unified Field.

-When we match ourselves to Organism Earth, we find there is nothing that we do that it does not do except speak through word/stories (abstracts).

29. How much of our ability to sense and feel do we inherit from nature?

Answer: All of it except our inaccurate or nature-disconnecting stories.

30. What is the relationship between our increasing personal social and environmental problems?

Answer: They all are symptoms or functions of our inaccurate or nature-disconnecting stories.

Go to a natural area and find examples there of the two most attractive of the five answers above.

31. What makes Natural Attraction Ecology be a pure and whole, rather than limited, objective science?

Answer: Its essence is its attraction to accepting 54 natural attraction sensations as being facts that are as real as material facts.

32. Where do we physically live in the solar system?

Answer: We live in, not on, Planet Earth, under its atmosphere and flying life and in its biosphere.

33. What is time in nature?

Answer: The immediate moment.

34. How do we know if Nature is alive? (Same as #28)

Answer:

-In any given moment all things in the Universe are identical as natural attraction manifesting itself as the Universe. For this reason, if any one thing is alive in a moment, all things are also alive. Any moment that a person knows they are alive means Nature and the Universe are alive including the Unified Field.

-When we match ourselves to Nature, we find there is nothing that we do that it does not do except speak through words.

35. What is natural homeostasis in nature and how does it work?

Answer: Homeostasis is all things being in diversity and balance to most adequately support life in ways that all things belong. It is a compromise between being attracted to the Unified Field and to new attractions that arise.

Go to a natural area and find examples there of the two most attractive of the five answers above.

36. Why do we excessively conquer or exploit nature?

Answer: We believe and are attracted to gaining satisfaction from harmful, nature-disconnected stories.

37. Why is it best to identify nature as a dance?

Answer: To avoid thinking of parts of nature's resonances and vibrations as being repulsion/repulsive or dualities.

38. Does duality exist in nature?

Answer: Only in human nature-disconnected stories and acting from them

39. What is the difference between nature and life?

Answer: The whole universe is alive; they are identical.

40. What is NNIAAL?

Answer: An acronym to help us immediately acknowledge and relate to nature as, as the truths of it **n**amelessness, immediate, intelligence, aliveness, attractiveness as a Unified Field and Planetary form of love.

Go to a natural area and find examples there of the two most attractive of the five answers above.

41. Are the Higgs Boson and NNIAAL identical?

Answer: Yes.

42. What is an Earth Avatar?

Answer: A person who knows themselves to be and Groks and acts as a personification of Earth's Unified Field.

43. What is the difference between nature and the Standard Universe?

Answer: There is none.

44. In our Other Body natural attraction Dance when is nature repulsive?

Answer: When it expresses itself in humanity as nature-disconnected stories and acts.

45. In the macro Natural Attraction Dance of the Unified Field, what are the names given to its micro pulsating resonances and fluctuations?

Answer: The attractive music, rhythm, vibrance, and sway of the dance.

> **Go to a natural area and find examples there of the two most attractive of the five answers above.**

46. What is the purpose and contribution of "Validating?"

Answer: Being consciously aware of the personal truth and global veracity of a fact or experience.

47. What is the significance of "Tropicmaking."

Answer: Recognizing that our attractions tend to build artificial Tropic-like replacements in the tropics, stories that become excessive and out of synch with non-tropical areas.

48. Why are webstrings often subconscious and how do we bring them into awareness?

Answer: Because they are discomforting when we are conscious of them after they have been injured or rejected by being bound to nature-disconnected stories.

49. How many natural senses do we have? Can we think with them? How?

Answer: 54 and we can think with them by energizing them into our consciousness when they are safely connected to and supported in natural areas.

50. How many natural senses can you name that you can know and learn from?

Answer: 54 when I read them from a document.

Go to a natural area and find examples there of the two most attractive of the five answers above.

51. How does nature within you know how to relate responsibly to nature in others and the environment?

Answer: By expressing and acting from its genetic attraction makeup that it shares with all other things.

52. If life has a purpose or love, what is it?

Answer: To support life.

53. Is our innate ability to sense and feel of, by and from nature?

Answer: Yes, it is how we are in communication with nature and vice versa through the varying intensity of 54 diverse natural attractions.

54. Do our formal education or leaders competently address the above questions?

Answer: They are aware of them because they are them but profits from tropicmaking have changed them into attachments to stories that omit, explore or conquer nature and produce the stress of discord in and around us.

55. Why don't our cultural stories and dollar bills generate self-correcting diversity, cooperation, balance, purity, and beauty, like nature works?

Answer: On average, over 99 percent of the way we learn to think and feel is disconnected from and out of tune with these attraction properties of the Unified Field.

Go to a natural area and find examples there of the two most attractive of the five answers above.

56. Do our unsolvable problems result from the difference between how we think and how nature's Unified Field works?

Answer: Yes, because we can think in abstract story attachments that omit or are scientific falsehoods about nature.

57. In nature, why don't two plus two equal four?

Answer: Unified Field attractions are continually changing as it grows more attractive time and space aliveness moment by moment, so "1" is never the same. One is the Unified Field that is everywhere, so zero does not exist either, since nowhere is there nothing, except in the nature-estranged way we learn to think.

58. Where in nature do you find life abstracted?

Answer: In our solar system in Planet Earth in Humanity's sense of Literacy as far as we know to date.

59. What in Nature is not conscious on some level?

Answer: Attraction is conscious of what it is attracted to. This means some degree of consciousness is always everywhere.

60. What parts of Nature is prejudiced against Nature?

Answer: Attachments to stories that misrepresent us, disconnecting us from the Unified Field.

Go to a natural area and find examples there of the two most attractive of the five answers above.

61. What is the relationship between natural attractions and consciousness?

Answer: Attraction is the Unified Field attracted to be conscious of what it is attracted to. The sense of Consciousness is attractive but does not, on its own, have a direction other than its attraction to be more conscious.

62. Is nature a form of perfection people can achieve and if so, how.

Answer: We are born as personifications of nature's perfection and can learn via ECHN to think and act as Nature works by consciously connecting our 54 natural senses to Nature, backyard or backcountry.

63. What is the element that produces environmentally and socially destructive economic relationships?

Answer: Attachments to Nature-disconnecting stories.

64. Since we are part of Nature, what is the significant difference that makes us destroy the

environment while everything else in Nature usually strengthens it?

Answer: Attachments to Nature-disconnecting stories.

65. How can we restore to our thinking the missing 49 sensory intelligences that contemporary society has buried in our subconscious?

Answer: Engage in 170 different ECHN activities that help us make time and space for the Unified Field to correct and restore itself in and as us.

Go to a natural area and find examples there of the two most attractive of the five answers above.

66. Why does modern society often identify a person's love of nature as "escapist recreation and fuzzy thinking" rather than "peaceful, reasonable and cooperative re-creation?"

Answer: Our thinking is educated and rewarded to attach to nature-conquering Tropicmaking stories.

67. What are nine additional words that mean the same thing as "natural attraction"?

Answer: Essence, Love, Life, To Be, Consciousness, Truth, Unified Field, Big Bang (and God if God is Love).

68. What is the essential contribution of the Unified Field Equation?

Answer: It is a cultural singularity, a “nine-leg” blueprint of words that help us re-connect our 54 natural senses to the wisdom of any moment's Unified Field time and space.

69. What is the value of identifying and relating through singularities?

Answer: When they are consciously connected to the Unified Field they are decisive, free of and immune to dualistic meanings and attachments that are confusing or detrimental.

70. What unique value does Scientific methodology, the Big Bang, mathematical logic and the Unified Field have in common?

Answer: They are each an established and reasonable logic sequence of natural attraction events or concepts that we are attached to and whose stories we can act out to increase well-being.

Go to a natural area and find examples there of the two most attractive of the five

answers above.

71. What are the benefits of recognizing that God is love and is found everywhere?

Answer: It makes God identical with, rather than separate from, the Unified Field and its benefits to the Web of Life and Societies in conflict.

72. What is the contribution of the GreenWave approach to the Web-of-Life model?

Answer: At will, it directly connects a person to the unifying attraction powers of the Unified Field in and around them and others.

73. Are the effects of applying scientific facts an accurate way of measuring them?

Answer: Yes, the effects are just as much reasonable facts as is anything else.

74. How does empirical thinking show that the Universe is alive?

Answer: In any given moment all things in the Universe are identical as natural attraction manifesting itself as the Universe. For this reason, if any one thing is alive in a moment, all things are also alive.

75. What sense makes sense to the other 53 senses by Validating experiences?

Answer: The sense of reason when it is conscious of them.

Go to a natural area and find examples there of the two most attractive of the five answers above.

76. Why does Scientific Methodology omit the contribution of sensory subjectivity?

Answer: Sensory "subjectivity" is Nature always changing, so it does not offer lasting standard conditions of pressure, temperature, motion, energy, attraction, intent, space, motivation or time that Science needs to measure it as a singularity. This omission deprives objective Science and Technology of the means to come into balance. It is remedied by scientific methodology including 54 natural attraction sensations in natural areas as facts of life that are part of any equation or decision. Then objective science becomes, or is also whole life science.

77. What is the value of attraction at the beginning of time, or before, being conscious of its attraction to support life?

Answer: It is the authority or resource that guides attraction to transform and organize itself moment-after-moment (m-m) appropriately.

78. How does a yellow zipper relate to a blue and orange sided piece of paper?

Answer: It transformatively zips together and intermingles or blends nature-disconnected stories on the orange side to support the life of true-blue Nature instead.

79. What is SUNEH-NNIAAL 54?

Answer: An acronym that helps us remember some innate, whole-life parts of Nature to help us become aware of them at will. Science, Universe, Nature, Earth Humanity is Now, Nameless, Intelligent, Alive, Attraction-love-54, GreenWave-54, webstring senses.

80. What is CRL?

Answer: CRL is the critical point source of any whole life moment when the senses of Consciousness and Reason are challenged to move into a nature-connected Literate-Story that supports rather than separates or conquers the life of Nature's.

Appendix E: Tree-ness

"The Field… the sole governing agent of each particle of matter."- Albert Einstein

"There is a field… and I'll meet you there." -Rumi

Albert Einstein depicted the ***Unified Field Theory*** as "*the fundamental forces of physics between elementary particles into a single theoretical framework.*" This is interpreted to mean the field unifies together all living beings, matter, particles, and so forth.

Religion, race, culture, creed, educational background, social-economic status, and so forth are "labels" for stories in our dialect world. These labels are merely words, customs that detach us from each other. In the Unified Field we are one, and regardless of how different we may seem to one another, we are equally balanced parts of a whole.

Rumi stated: "*You are not a drop in the ocean. You are the entire ocean in a drop.*" This is the foundation for humankind, and all life, even on a molecular level – ***EVERYTHING is connected.*** Every idea, action, and response have an implication on a grander scale, which is more expansive then we could ever envision. We no longer "***go to nature to be soothed and healed, and to have my senses put in order.***" (*John Burroughs*) We have replaced this with "escapism." We go to escape the injury of the everyday moment to moment life, never experiencing the simplicity of our breath in those moments. In this very moment, and in the next one, it is critical, more than any time in history, to embrace this unification and recognizing we are all responsible for not only ourselves but each another. We are all essential branches of the same huge lovely "Tree of Life."

> Many individuals do not realize that when a caterpillar goes into its cocoon to transform into a butterfly, it is part of the Tree of Life wisely metamorphizing. We frequently learn about the distinct stages of the caterpillar to butterfly change. The caterpillar, larva, eats leaves making itself plumper and longer through a progression of sheds where it molts its skin. One day, the caterpillar quits eating and hangs upside down from a twig or leaf and spins itself into a protective casing, a glossy chrysalis. Starting here, the caterpillar transforms itself into the beautiful butterfly many people have grown to appreciate and love.
>
> What happens inside the cocoon, is like a mad scientist lab experiment! The caterpillar is attracted to digest itself, releasing enzymes to dissolve all of its tissue. If you somehow managed to cut open a cocoon at the correct time, it would overflow like a slimy soup. So how is it this ooze turns into a butterfly?

> By all appearances it is dead. The process inside the cocoon is scientifically complicated. Once a caterpillar has deteriorated all of its tissues, aside from the imaginal discs, those discs are attracted to use the protein-rich soup surrounding them to fuel the active cell division required to frame the wings, antennae, legs, eyes, genitals and the various highlights of an adult butterfly or moth. When in this stage, disturbing the process can result in stopping the transformation process, resulting in its demise.
>
> Additional Information and images of the transformation process may be seen at http://www.wormspit.com/pernyi.htm

The Unified Field is the ***Field of Life*** that is being Us moment to moment (m/m). A significant portion of what numerous individuals encounter m/m is devoid of unity. Along these lines, we resemble hungry baby birds, our bills open and squawking for our Mother to come back with the nourishment we require. In comparison, humankind waits to be fed by its innovative technology and other "pacifiers" which dull its GreenWave-54-senses. Nevertheless, if we feel nature in and around us, we can sense life is always present. We can sense that when we are that ravenous fledgling, we are satisfying a need for sustenance, survival, love and so forth. While sitting tight for our Mother to bring us nourishment, we are still supported by the nest she painstakingly built for us. In this home, we are protected, secure, warm, and watched over. This same nest is recycled by nature when it is no longer our home. Our Mother will not miss feeding us since we are interconnected and her senses guide her tenderly back to us each time we require her. In the far-fetched occasion, she succumbs to misfortune, other beings in nature frequently assist in ensuring survival. This is possible because Webstrings continuously flow m/m weaving themselves into the GreenWave.

Since we have become attached to the story world and emotional distress (senses 25-27), we have been shown things die, and change, and are transient. However, each living being, at the deepest level is the Unified Field, and that is the reason we encounter this change. In light of the fact the fundamental level of who we are is perpetually growing and collapsing back in on itself is a massive feedback loop making the story world come forward and changing before us as an excellent, multifaceted, and grand motion picture played only for us. This is devoid of understanding, once more, of unity and the Webstrings we share as a whole.

As the saying goes, "time flies when you're having fun." This is relative to discoveries and statements Einstein made because of the self-evident fact when we encounter life with great love and energy it is bringing us closer to the Unified Field. The rate of time accelerates according to our perception. The "tipping point" is the GreenWave peak where we are in the now moment. It is then we encounter the full effect of a moment devoid of time-space. Between our thoughts and language is the GreenWave-54 Unified Field. When we become the silence, we experience we turn out to be more than the Unified Field. We become universal. Our nature negatives return to balance, and we encounter freedom, pureness, and immense love.

The Earth created a perfect and glorious garden that was energetically alive. As a result, the Earth's awareness utilized the Unified Field of information, along with similar components Einstein talked about. When we experience change on Earth, it is because we are encountering the ebb and flow of the Earths' sensory aliveness. She is creating in amicability with the solar, water, moon cycles, and so forth. Every human is encountering our cycles of expansion and experience at any given moment.

The ***Tree of Life*** is a natural being, a blueprint, created to ensure our survival. Bring into your mind's awareness a tree you have gone by or one that is outside your window. Envision all renditions of the tree, from birth to the one you see now, alongside the future "death." All are inside one space and time in the Unified Field. By recognizing the Tree of Life never indeed appears, dies, or moves on its own, it is only our view of reality which gives the tree the appearance it is evolving. What is occurring in the Unified Field is that all time, moment to moment, is continually in motion and all movement is the result of our consciousness of it being in action. The Unified Field, as it expands and contracts all other things around it, is perceived from a material level. We are enjoying the creation of evolution as new levels of the Unified Field become visible to us as a result of our shift in our sensory awareness. Since, as a whole, we all share a universal consensus of reality (generally) physical time emerges as a story in humanity. In the event we were ever in a time-space void, time ceases to exist because there are no objects to determine motion or cycles of living beings. Therefore, we would be in a state of perpetual stillness and timelessness, the GreenWave Unified Field!

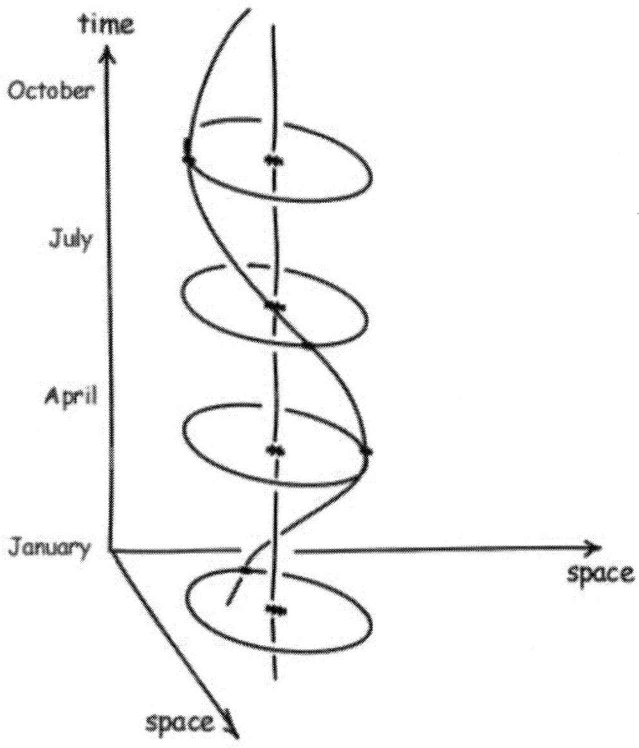

Figure 1: Earth Orbiting Sun (3-D)

In ***Figure 1*** above the Sun is represented by the marking in the middle. The position of the Earth for the months displayed demonstrate their position for those months. Does the formation (line connecting the point of time/space for each month) remind you of anything? If so, what?

Additional (Optional Reading):

Study: A Bose-Einstein condensate (BEC). What is it? Why does it matter?

Article on the first BEC Created In January 2017:
http://www.laserfocusworld.com/articles/2017/01/bose-einstein-condensate-successfully-created-in-space-for-the-first-time.html

EcoArt-Therapy Activity: Draw Yourself as Tree-ness

> **"Humankind is in desperate need of a shift from a mentality of competition to one of cooperation." -Stacey S. Mallory**

1. Draw yourself as a section of a tree. Leave enough space to draw the upper branches and roots.
2. Now transform your tree into tree-ness:

    - Draw the ground for your tree.
    - Draw roots of various sizes going into the ground to demonstrate senses you are attracted to moment/moment (m/m) such as feeling nourished, grounded, etc.
    - Draw branches to demonstrate attractions to senses you can sense around you such as the sky, air, color, sound, etc. Draw leaves, flowers or fruit to express other attractions you may have. Don't be afraid to draw abstractly! Put fruit or symbols of how you sense something inside of leaves or flowers.

Reflective Questions

1. What qualities do you like about the tree?
2. Try to go back to the tree on another day. How is the tree different? How are your GreenWave-54 attractions different?
3. Write down what GreenWave-54 feelings come to you as you draw yourself as a tree into tree-ness. How do the two differ?
4. Are some roots, branches, leaves, and so forth larger or smaller than others? If so, does this represent anything for you?
5. Is there anything in your drawing you did not want to add, but did? If so, what and why?

Appendix F: References

A new Copernicun revolution. (2012). *Journal of Organic Psychology and Natural Attraction Ecology, 2. Retrieved from* *http://www.ecopsych.com/journalcopernicus.html*

Alban, D. (2016) *Epigenetics: How you can change Your Genes and Change Your Life.* Retrieved from *http://reset.me/story/epigenetics-how-you-can-change-your-genes-and-change-your-life/*

Albert Einstein's Unified Field equation. (2014-2016). *Journal of Organic Psychology and Natural Attraction Ecology, 2.* Retrieved from *http://www.ecopsych.com/einsteinstart.html*

Block, J. (2016) *First in Nation Lawsuit Over Climate Change.* Retrieved from *http://www.clf.org/newsroom/clf-files-lawsuit-against-exxonmobil/*

Brown, N. (1990). *Love's body.* University of California Press.

Cohen, M. J. (1983). Prejudice Against Nature. *Cobblesmith.* Retrieved from *http://www.ecopsych.com/prejudicebigotry.html*

Cohen, M. J. (1993). *The training ground of a nature-connected expert.* (2014). Retrieved from *http://www.ecopsych.com/mjcohen.html*

Cohen, M. J. (1994) *The global wellness and unity activity.* Retrieved from *http://www.ecopsych.com/amental.html*

Cohen, M. J. (1995). Education and counseling with nature: A greening of psychotherapy. *The Interspsych Newsletter, 2*(4). Retrieved from *http://www.ecopsych.com/counseling.html*

Cohen, M. J. (2007a). *Thinking and feeling, and relating through the joy of nature's perfection.* Retrieved from *http://www.ecopsych.com/naturepath.html*

Cohen, M. J. (2007b). *Who am I? Who or what is your natural self?* Retrieved from *http://www.ecopsych.com/thesisquote6.html*

Cohen, M. J. (2008). *Educating, counseling, and healing with nature.* Illumina. Retrieved from *http://www.ecopsych.com/ksanity.html*

Cohen, M. J. (2009a). *How to transform destructive thinking into constructive relationships.* Retrieved from *http://www.ecopsych.com/transformation.html*

Cohen, M. J. (2011). *The anatomy of institutions.* Retrieved from *http://www.ecopsych.com/journalinstitution.html*

Cohen, M. J. (2012). *The Slime Mold Alternative.* Retrieved from *http://www.ecopsych.com/journalslimemold.html*

Cohen, M. J. (2013). *The great sensory equation dance.* Retrieved from http://www.ecopsych.com/journalgut.html

Cohen, M. J. (2014). *Benefit from consciously registering your fifty-four natural senses.* Retrieved from http://www.ecopsych.com/insight53senses.html

Cohen, M. J. (2015). *A Survey of Nature-connected learning participants.* Retrieved from http://www.ecopsych.com/survey.html

Cohen, M. J. (2016). *Maverick Genius Walk.* Retrieved from http://www.ecopsych.com/MAKESENSEWALK.docx

Cohen, M. J. (2017) *The Missing Element in Unsolvable Problems: Attraction is Conscious of What it is Attracted to.* Retrieved from http://www.ecopsych.com/GREENWAVE.docx

Dewey, J. (1934). Individual psychology and education. *The Philosopher, 2.* Retrieved http://www.ascd.org/ASCD/pdf/journals/ed_update/eu201207_infographic.pdf

Doherty, T. J. (2010). Michael Cohen: Ecopsychology interview. *Ecopsychology Journal, 2.* Retrieved from http://www.ecopsych.com/ecopsychologyjournal.html

Einstein, A., Cohen, M.J. (2014-2016)). Albert Einstein's Unified Field equation. *Journal of Organic Psychology and Natural Attraction Ecology, 2.* Retrieved from http://www.ecopsych.com/einsteinstart.html

Fishman, K. (1854). Chief Seattle's speech. *Wildwood Survival.* Retrieved from http://www.wildwoodsurvival.com/wildernessmind/chiefseattle.html

Grange [#966]. (2015). Resolutions. Retrieved from http://www.sjigrange.wordpress.com/resolutions

Green, B. (2013). How the Higgs Boson was found. *Smithsonian Magazine.* Retrieved from *http://www.smithsonianmag.com/science-nature/how-the-higgs-boson-was-f und 4723520/?cmd=ChdjYS1wdWItMjY0NDQyNTI0NTE5MDk0Nw&page=3*

Hoke, P. (2015) Maverick Genius at Work. Retrieved from http://www.ecopsych.com/maverick-genius

Holthaus, E. (2016) *Kids Suing Government Over Climate Change.* Retrieved from *http://www.slate.com/articles/health_and_science/science/2016/11/the_kids_lawsuit_over_climate_change_is_our_best_hope_now.html*

Kluger, J. (2014). *The sixth great extinction.* Time Magazine. Retrieved from *http://time.com/3035872/sixth-great-extinction/*

Milgram, (1974). Obedience to authority: An experimental view. Harpercollins. Retrieved from *https://en.wikipedia.org/wiki/Milgram_experiment*

Murchie, G. (1999) Seven Mysteries of Life, Mariner Publishing.

Pascal, B. (1995). *Pensées.* (p. 312). Oxford University Press, USA.: Penguin Books Retrieved from *http://www.naturalchild.org/jan_hunt/babyspeaks.html*

Singer, J. (n.d.). Ode to Positive Constructive Daydreaming.

Skinner, B.F. (1971) *Beyond Freedom & Dignity,* Pelican Books.

Robinson, T.N. (2007) *Effects of Fast Food Branding on Young Children's Taste Preferences* Journal of the American Medical Association Retrieved from *http://jamanetwork.com/journals/jamapediatrics/fullarticle/570933*

International Astronomical Union. (2015). *Universe is dying.* Retrieved from *http://news.discovery.com/space/galaxies/universe-is-dying-galactic-survey-shows-150810.htm*

Natural attraction ecology. (2003). Retrieved from *http://www.naturalattractionecology.com/?page_id=30*

Our living universe: Who is the boss of you? (2014). Retrieved from *http://www.ecopsych.com/universealive.html*

Peak fact: Whole life self-evidence in action. (2010). *Journal of Organic Psychology and Natural Attraction Ecology, 2.* Retrieved from *http://www.ecopsych.com/journalpeak.html*

Project NatureConnect (PNC). (1997). *Reconnecting with Nature,* EcoPress. Retrieved from *http://www.ecopsych.com/insight53senses.html*

The hidden organic remedy: Nature as a higher power. (2013). *Journal of Organic Psychology and Natural Attraction Ecology, 1.* Retrieved from *http://www.ecopsych.com/nhpbook.html*

The impossible dream: We ask you to be a part of it. (2011-2013). *Journal of Organic Psychology and Natural Attraction Ecology, 1.* Retrieved from *http://www.ecopsych.com/journalwarranty.html*

The magic of something from nothing. (2012). Journal of Organic Psychology and Natural Attraction Ecology, 2. Retrieved from *http://www.ecopsych.com/journalessence.html*

The National Grange. (1874, February 11). The declaration of purposes of the National Grange, Retrieved from *http://www.oocities.org/cannongrange/declaration_purposes.html*

The state of planet earth and us. (2001). Retrieved from *http://www.ecopsych.com/zombie2.html*

Thinking and learning with all nine legs. (2011). Retrieved from *http://www.ecopsych.com/nineleg.html*

Who, what or when is the acronym NNIAAL? (2013). Retrieved from *http://www.ecopsych.com/earthstories101.html*

Searls, D. (2009). The Journal of Henry David Thoreau, 1837-1861 December 6 entry, New York Review Books Classics. Retrieved from *https://en.wikiquote.org/wiki/Henry_David_Thoreau*

Made in the USA
Columbia, SC
17 January 2018